Ultrafine-Grained Metals

Special Issue Editor
Heinz Werner Höppel

MDPI • Basel • Beijing • Wuhan • Barcelona • Belgrade

MDPI

Special Issue Editor
Heinz Werner Höppel
Friedrich-Alexander-University Erlangen-Nürnberg FAU
Germany

Editorial Office
MDPI AG
St. Alban-Anlage 66
Basel, Switzerland

This edition is a reprint of the Special Issue published online in the open access journal *Metals* (ISSN 2075-4701) from 2014–2015 (available at: http://www.mdpi.com/journal/metals/special_issues/ultrafine-grained-metals).

For citation purposes, cite each article independently as indicated on the article page online and as indicated below:

Author 1; Author 2. Article title. *Journal Name* **Year**, *Article number*, page range.

First Edition 2017

ISBN 978-3-03842-524-3 (Pbk)
ISBN 978-3-03842-525-0 (PDF)

Table of Contents

Chapter 1: Editorial

Chapter 2: UFG Titanium Alloys

Chapter 3: UFG Iron and Steels

Chapter 4: UFG Aluminium Alloys

Chapter 5: UFG Copper Alloys

About the Special Issue Editor

Heinz Werner Höppel born in 1968, studied Materials Science and Engineering at the University of Erlangen-Nürnberg. From 1994 until 1997 he worked as a research associate at the Department of Materials Science and Engineering, Institute I, General Materials Properties, on the field of hydroabrasive wear resistance and damage mechanisms of different hard coatings. After he has finalized his PhD-thesis, he took over a senior scientist position at the Institute I. In 2015 Dr. Höppel competed his habilitation thesis on "Monotonic and cyclic deformation behaviour of very fine grained metals: Microstructural mechanisms and mechanical properties". Currently he is the head of the laboratories for Mechanical Testing and for Nanometals. His main scientific interests are on the fields of fatigue and cyclic plasticity, fretting fatigue, tribological behaviour and in particular ultrafine-grained materials.

Preface to "Ultrafine-Grained Metals"

Written by designated researchers this book provides a compilation of current results of research activities in the field of ultrafine-grained metals. The different papers address not only recent advances in understanding relevant processes in severe plastic deformation, they also span over manifold material systems, like iron and steels and also titanium-, aluminum- and copper-alloys. Furthermore, not only results on (the outstanding) mechanical properties of UFG materials are reported in detail but also the effect of UFG-microstructures on the electrical conductivity and the pitting corrosion behavior are shown. Thus, this book provides a thematically wide survey on the ongoing research activities in the field of ultrafine-grained metals and is believed to be gainful for those scientist who entered the field of UFG metals for the first time and for those who are already very well experienced in the field.

Heinz Werner Höppel
Special Issue Editor

Chapter 1:
Editorial

metals

MDPI

Editorial
Ultrafine-Grained Metals

Heinz Werner Höppel

Department of Materials Science and Engineering, Institute I: General Materials Properties, Friedrich-Alexander Universität Erlangen-Nürnberg, Martensstr. 5, 91058 Erlangen, Germany; hwe.hoeppel@fau.de; Tel.: +49-9131-8527503

Received: 10 December 2015; Accepted: 10 December 2015; Published: 16 December 2015

Ultrafine-grained (UFG) metallic materials are at the cutting edge of modern materials science as they exhibit outstanding properties which make them very interesting for prospective structural or functional engineering applications. Due to the progress in severe plastic deformation techniques during the last decade, ultrafine-grained microstructures are no longer only restricted to easy to deform single-phase materials, but can also be introduced in complex and hard to deform alloys of technological relevance. Nowadays it is of course clear that not only the hardening effect by the well-known Hall-Petch law is of high importance in UFG materials, but many other issues also come into play. This Special Issue on ultrafine-grained metals covers a broad range of research activities in that field. Fifteen articles have been selected for this issue addressing manifold topics such as new developments in severe plastic deformation techniques, advances in modeling and simulation of the severe plastic deformation processes, mechanical properties under monotonic and cyclic loading of homogenous and graded UFG structures and, related to that, dominating deformation mechanisms in UFG materials. Furthermore, advances and strategies for high conductivity materials by introducing UFG structures; the correlation between severe plastic deformation parameters and the resulting materials properties; and peculiarities in the corrosion behavior of UFG materials, have been addressed.

The 15 articles reflect, on the one hand, the variety of state of the art research activities in the field and, on the other hand, form the spearhead of current advances. The articles are briefly summarized as follows:

- Murashkin *et al.* [1] studied the influence of an ECAP-conform process on the mechanical properties and the electrical conductivity of an AA6101 alloy. They found out that both the electrical conductivity and the mechanical strength can be significantly increased by this process when the data are compared to the T6 or T81-CG counterparts.
- Sajadifar and Yapic [2] modified a Johnson-Cook model in order to predictively describe the flow behavior of UFG titanium at elevated temperatures, which also aims to describe high-temperature forming processes.
- Murdoch *et al.* [3] succeeded in introducing a graded grain structure by applying a surface mechanical attrition treatment (SMAT) to a bcc iron plate at ambient and cryogenic temperatures.
- Rufing *et al.* [4] showed that a high-pressure torsion (HPT) process applied to an SAE1045 steel results in a significantly enhanced endurance limit. They also showed that the fatigue crack initiation mechanisms were changed by the severe plastic deformation.
- Another contribution in the field of fatigue, by Köhler *et al.* [5], deals with the influence of an ECAP-process on unreinforced and particulate reinforced AA2017 on the fatigue crack growth behavior.
- Lee *et al.* [6] investigated the potential of multi-pass caliber rolling of Ti6Al4V to obtain an UFG microstructure in larger quantities. It was also found that the material in the UFG state exhibits superplastic behavior.
- Altenberger *et al.* [7] addressed the increasing demand in the industry for highly conductive high strength copper alloys. New strategies are shown to achieve a high strength paired with good

conductivity by introducing an UFG microstructure. Furthermore, the role of precipitates for thermal stability is investigated and promising concepts and alloy systems for the future are proposed and discussed.

○ Ruppert *et al.* [8] subjected the austenitic stainless steels X4CrNi18-12 and X8CrMnNi19-6-3 to the ARB-process and obtained rather high yield strengths of more than 1.25 GPa paired with a high ductility. It was shown that micro-twinning comes into play.

○ High pressure torsion was used by Krämer *et al.* [9] to produce bulk metallic glass by severe plastic deformation. They consolidated Zr-based metallic glass powder and deform it further to weld the powder particles together.

○ Odnobokova *et al.* [10] studied the effect of large strain cold rolling at ambient temperature on a 304 L stainless steel. They found that deformation twinning followed by micro-shear banding and martensitic transformation promoted the development of a nanocrystalline structure with an extraordinary high yield strength of 1.6 GPa.

○ Murashkin *et al.* [11] reported on the monotonic and cyclic properties of an ultrafine-grained Al 6061 alloy processed by high-pressure torsion. A significant improvement of the monotonic strength as well as the fatigue strength is reported.

○ Ebrahimi *et al.* [12] investigated a rather new SPD-process, called equal channel forward extrusion, and obtained rather similar properties and microstructures as obtained by other severe plastic deformation methods.

○ Semenova *et al.* [13] consolidated cryogenic milled Ti-powder by high pressure torsion and investigated the obtained microstructures. Grains smaller than 40 nm have been obtained.

○ Nickel *et al.* [14] studied the effect of strain localization on pitting corrosion of an AlMgSi0.5 alloy. It was found that more pits emerge in shear bands, but the pit depth is reduced significantly. Moreover, stable pitting of shear bands results in less positive potentials compared to adjacent microstructures.

○ Ma *et al.* [15] applied multi-pass equal-channel angular pressing (EACP) to produce ultrafine-grained Cu-0.2 wt % Mg alloy contact wires. They showed that Cu-Mg alloy after multi-pass ECAP at 473 K exhibit a high strength paired with satisfactory conductivity.

In summary, these 15 articles excellently highlight the diversity of research in the field of ultrafine-grained metals and offers a snapshot of current activities in the field. The obtained results clearly show, on the one hand, the high potential of UFG materials; and on the other hand, although the first research activities in the field were started in the early 1980s, it becomes obvious that there are still plenty of open topics to be investigated. As stated in the beginning: The potential of ultrafine-grained materials is not only related to a high strength according to the Hall-Petch law but also to many other issues that come up or change with grain size.

Conflicts of Interest: The authors declare no conflict of interest.

References

1. Murashkin, M.; Medvedev, A.; Kazykhanov, V.; Krokhin, A.; Raab, G.; Enikeev, N.; Valiev, R.Z. Enhanced Mechanical Properties and Electrical Conductivity in Ultrafine-Grained Al 6101 Alloy Processed via ECAP-Conform. *Metals* **2015**, *5*, 2148–2164. [CrossRef]
2. Sajadifar, S.V.; Yapici, G.G. High Temperature Flow Response Modeling of Ultra-Fine Grained Titanium. *Metals* **2015**, *5*, 1315–1327. [CrossRef]
3. Murdoch, H.A.; Darling, K.A.; Roberts, A.J.; Kecskes, L. Mechanical Behavior of Ultrafine Gradient Grain Structures Produced via Ambient and Cryogenic Surface Mechanical Attrition Treatment in Iron. *Metals* **2015**, *5*, 976–985. [CrossRef]
4. Ruffing, C.; Kobler, A.; Courtois-Manara, E.; Prang, R.; Kübel, C.; Ivanisenko, Y.; Kerscher, E. Fatigue Behavior of Ultrafine-Grained Medium Carbon Steel with Different Carbide Morphologies Processed by High Pressure Torsion. *Metals* **2015**, *5*, 891–909. [CrossRef]

5. Köhler, L.; Hockauf, K.; Lampke, T. Influence of Particulate Reinforcement and Equal-Channel Angular Pressing on Fatigue Crack Growth of an Aluminum Alloy. *Metals* **2015**, *5*, 790–801. [CrossRef]

6. Lee, T.; Shih, D.S.; Lee, Y.; Lee, C.S. Manufacturing Ultrafine-Grained Ti-6Al-4V Bulk Rod Using Multi-Pass Caliber-Rolling. *Metals* **2015**, *5*, 777–789. [CrossRef]

7. Altenberger, I.; Kuhn, H.A.; Gholami, M.; Mhaede, M.; Wagner, L. Ultrafine-Grained Precipitation Hardened Copper Alloys by Swaging or Accumulative Roll Bonding. *Metals* **2015**, *5*, 763–776. [CrossRef]

8. Ruppert, M.; Freund, L.P.; Wenzl, T.; Höppel, H.W.; Göken, M. Ultrafine-Grained Austenitic Stainless Steels X4CrNi18-12 and X8CrMnNi19-6-3 Produced by Accumulative Roll Bonding. *Metals* **2015**, *5*, 730–742. [CrossRef]

9. Krämer, L.; Kormout, K.S.; Setman, D.; Champion, Y.; Pippan, R. Production of Bulk Metallic Glasses by Severe Plastic Deformation. *Metals* **2015**, *5*, 720–729. [CrossRef]

10. Odnobokova, M.; Belyakov, A.; Kaibyshev, R. Development of Nanocrystalline 304 L Stainless Steel by Large Strain Cold Working. *Metals* **2015**, *5*, 656–668. [CrossRef]

11. Murashkin, M.; Sabirov, I.; Prosvirnin, D.; Ovid'ko, I.; Terentiev, V.; Valiev, R.; Dobatkin, S. Fatigue Behavior of an Ultrafine-Grained Al-Mg-Si Alloy Processed by High-Pressure Torsion. *Metals* **2015**, *5*, 578–590. [CrossRef]

12. Ebrahimi, M.; Djavanroodi, F.; Tiji, S.A.N.; Gholipour, H.; Gode, C. Experimental Investigation of the Equal Channel Forward Extrusion Process. *Metals* **2015**, *5*, 471–483. [CrossRef]

13. Semenova, I.; Timokhina, I.; Islamgaliev, R.; Lavernia, E.; Valiev, R. Nanocrystalline Ti Produced by Cryomilling and Consolidation by Severe Plastic Deformation. *Metals* **2015**, *5*, 206–215. [CrossRef]

14. Nickel, D.; Dietrich, D.; Mehner, T.; Frint, P.; Spieler, D.; Lampke, T. Effect of Strain Localization on Pitting Corrosion of an AlMgSi0.5 Alloy. *Metals* **2015**, *5*, 172–191. [CrossRef]

15. Ma, A.; Zhu, C.; Chen, J.; Jiang, J.; Song, D.; Ni, S.; He, Q. Grain Refinement and High-Performance of Equal-Channel Angular Pressed Cu-Mg Alloy for Electrical Contact Wire. *Metals* **2014**, *4*, 586–596. [CrossRef]

metals

MDPI

Article

Experimental Investigation of the Equal Channel Forward Extrusion Process

Mahmoud Ebrahimi [1,*], Faramarz Djavanroodi [1,2], Sobhan Alah Nazari Tiji [1], Hamed Gholipour [1] and Ceren Gode [3]

[1]   Department of Mechanical Engineering, Iran University of Science and Technology, Tehran 16846-13114, Iran; fdjavanroodi@pmu.edu.sa (F.D.); s.nazari86@yahoo.com (N.T.); h_golipour@yahoo.com (H.G.)
[2]   Department of Mechanical Engineering, Prince Mohammad Bin Fahd University, Al Khobar 31952, Saudi Arabia
[3]   School of Denizli Vocational Technology, Program of Machine, Pamukkale University, Denizli 20100, Turkey; cgode@pau.edu.tr
*   Author to whom correspondence should be addressed; mebrahimi@iust.ac.ir; Tel.: +98-914-401-7268.

Academic Editor: Heinz Werner Höppel
Received: 3 February 2015; Accepted: 3 March 2015; Published: 16 March 2015

Abstract: Among all recognized severe plastic deformation techniques, a new method, called the equal channel forward extrusion process, has been experimentally studied. It has been shown that this method has similar characteristics to other severe plastic deformation methods, and the potential of this new method was examined on the mechanical properties of commercial pure aluminum. The results indicate that approximate 121%, 56%, and 84% enhancements, at the yield strength, ultimate tensile strength, and Vickers micro-hardness measurement are, respectively, achieved after the fourth pass, in comparison with the annealed condition. The results of drop weight impact test showed that the increment of 26% at the impact force, and also decreases of 32%, 15%, and 4% at the deflection, impulse, and absorbed energy, are respectively attained for the fourth pass when compared to the annealed condition. Furthermore, the electron backscatter diffraction examination revealed that the average grain size of the final pass is about 480 nm.

Keywords: SPD; ECFE; mechanical properties; impact behavior; grain size

---

1. Introduction

In the last decade, production and application of ultra-fine grain (UFG) and nano-structure (NS) metals and alloys have been deeply studied by researchers and scientists in the material science field [1,2]. These materials possess improved mechanical properties at the room temperature and enhanced superplastic behavior at higher temperatures [3,4]. In general, there are two main processing categories to fabricate UFG and NS materials, called bottom-up and top-down manners. In the bottom-up method, the UFG or NS materials are synthesized, atom-by-atom, and, also, in layer-by-layer arrangement, and that these samples possess small dimensions with porous structures, which are rarely appropriate for industrial applications. In the top-down approach, micro-structure (MS) materials at the industry scale have been altered to UFG, and even NS ones, using severe plastic deformation (SPD) techniques [1,5].

Up to now, numerous SPD methods have been proposed, experimented, and investigated in detail. Based on the geometry of the work-piece, SPD techniques can be divided into the three groups nominating bulk, sheet, and tube classifications. For group 1: equal channel angular pressing (ECAP) [1], high pressure torsion (HPT) [2], twist extrusion (TE) [6], accumulative back extrusion (ABE) [7]; for group 2: equal channel angular rolling (ECAR) [8], accumulative roll bonding (ARB) [9], constrained groove pressing (CGP) [10]; and for group 3: high pressure tube twisting (HPTT) [11], accumulative spin bonding (ASB) [12], tubular channel angular pressing (TCAP) [13] are the major examples. It should be pointed out that the principal rule of all SPD methods consists of imposing shear stress to the sample, increasing the dislocation density in deformed material, formation of dense dislocation walls and then low angle grain boundaries (LAGBs), and, finally, transformation of LAGBs into high angle grain boundaries (HAGBs) [1,14–17].

Recently, a novel SPD method called the equal channel forward extrusion (ECFE) process has been proposed by authors to fabricate UFG and NS bulk materials [18]. This research focuses on the capabilities of this new method. Hence, mechanical and microstructural investigations have been carried out to observe and compare the characteristics of this new method with the other SPD processes. For this aim, commercial pure aluminum billets have been ECFEed up to four passes and then the potential of this approach has been investigated by means of a tensile test, hardness measurements, an impact test, and microstructural observations.

2. Principle of the ECFE Process

The equal channel forward extrusion process is schematically represented in Figure 1a. As can be observed, the ECFE die consists of three major parts: inlet or entry channel, main deformation zone (MDZ), and outlet or exit channel. A sample with a rectangular cross-section is placed in the entry channel and then pushed by punch to pass through the entrance channel and enter the MDZ, where the material is subjected to intense shear stress. As the material passes through the MDZ, the cross-section of the sample incrementally expands in the width direction, and reduces in the length direction, simultaneously. Figure 1b represents the alteration of both the length and width of billets' cross-section to each other at the MDZ section, where the area of the rectangular sample remains constant during the process. It needs to be emphasized that there is no sample rotation during the process.

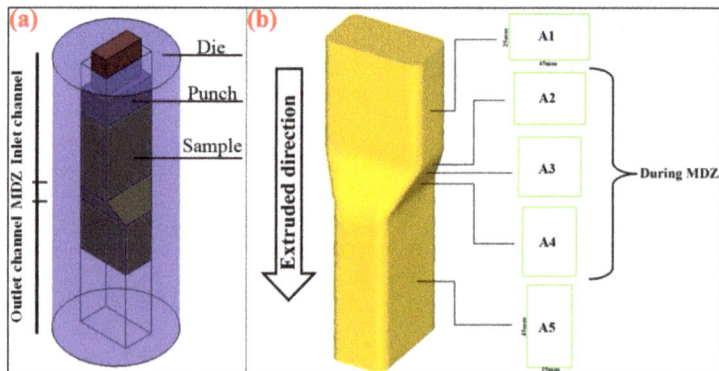

Figure 1. (a) The schematic representation of equal channel forward extrusion process and (b) the length to width alteration at the billet's cross-section in the MDZ section during the ECFE process where A1 = A2 = A3 = A4 = A5.

It is noted that the length to width ratio of the sample, and, also, the magnitude of MDZ's height are the two major parameters that influence the strain behavior, mechanical properties, and microstructural characteristics of the deformed materials.

3. Experimental Procedure

3.1. Materials

Commercial pure (CP) aluminum (Al1070) billets with the chemical composition of (in wt.%) 0.185% Fe, 0.09% Si, 0.012% Mg, 0.011% Zn, 0.008% Ti, 0.008% V, 0.004% Ni, 0.004% Cu, 0.003%

Mn, and Al as the balance, were prepared with the dimensions of 25 mm × 45 mm × 140 mm. Before extrusion, all the samples were annealed at 380 °C for 1.5 h and then cooled slowly in a furnace to room temperature [19,20]. This leads to a homogeneous and uniform structure with good ductility in all specimens before operation.

3.2. ECFE Die

The equal channel forward extrusion set-up, which includes die and punch parts, was designed and manufactured to the following specifications: (1) The die was made of 1.2510 tool steel (hardened up to 50 HRC) with the channel's cross-section being 25 mm × 45 mm in the three separate parts, namely entrance channel, MDZ, and exit channel; and (2) The punch was constructed with 1.2344 tool steel (hardened up to 55 HRC). Figure 2 displays the hydraulic press, ECFE die, and CP aluminum samples after extrusion up to the four passes. As can be seen, there is no considerable change in the dimensions of the billet samples after the process. The ECFE process was performed at ambient temperature. Molybdenum disulfide (MoS_2) was applied as a lubricant to reduce the frictional influence between the die and billet, and, also, the punch speed was equal to 2 mm/s during the operation.

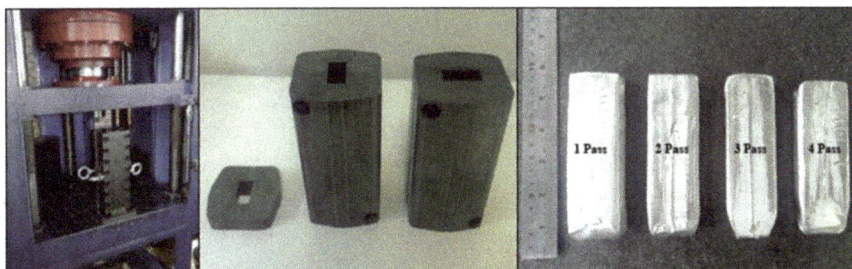

Figure 2. The hydraulic press, ECFE die parts (entrance channel, MDZ and exit channel), and CP aluminum billets after the ECFE process up to the four passes.

3.3. Microstructural Testing

Optical microscopy (OM) was applied by use of Clemex Vision PE software (Clemex, Denizli, Turkey) in accordance with the ASTM E112 to measure the average grain size of the annealed condition. Additionally, the classic Williamson-Hall technique of the X-ray diffraction (XRD, Philips, Tehran, Iran) patterns was employed to theoretically calculate the cell/sub-grain size of the deformed billets. The XRD analysis was conducted on the polished sections of the deformed samples in a Philips X-ray diffractometer, equipped with a graphite monochromator using $CuK\alpha$ radiation (1.541 Å) with an initial angle of 4°, step size of 0.05°, and step time of 3 s. Cell/sub-grain size and lattice distortion are two prominent crystalline imperfections of the materials. These imperfections lead to peak broadening in X-ray diffraction (XRD) patterns [21]. Cell/sub-grain size and lattice micro-strain can be calculated by measuring the deviation of the line profile from the perfect crystal diffraction. In the classic Williamson-Hall technique, cell/sub-grain size of material can be estimated using Equation (1) [22]:

$$\beta \cos \theta = \frac{k\lambda}{d} + f(\varepsilon) \sin \theta \tag{1}$$

$$\beta_{exp}^2 \cong \beta_{ins}^2 + \beta^2 \tag{2}$$

where, β is the integral breadth of the XRD profile, k is the shape factor (0.9), d is the cell/sub-grain size, λ is the wavelength, ε is the lattice strain, θ is the Bragg angle, and $f(\varepsilon)$ is a defined function. Furthermore, by considering that the experimental profile (β_{exp}) is the convolution of the instrumental profile (β_{ins}) and the intrinsic profiles (β), the intrinsic profile can be attained by unfolding the experimental profile via the Gaussian assumption, as is represented by Equation (2) [22,23]. Moreover, using Equation (1), a line can be fitted by plotting $\beta Cos\theta$ against $Sin\theta$ and the intercept gives the cell size. This line is known as Williamson-Hall graph. Thus, the above equation can be rewritten as:

$$
\begin{cases}
Y = \beta \cos\theta \\
X = Sin\theta \\
a = f(\varepsilon) \\
b = \frac{0.9\lambda}{d}
\end{cases}
\Rightarrow \beta \cos\theta = \frac{0.9\lambda}{d} + f(\varepsilon)\sin\theta \Rightarrow Y = b + aX
\tag{3}
$$

Then, both cell/sub-grain size and strain can be estimated from the y-intercept and the slope of the line, respectively. Figure 3a displays the XRD patterns of the Al1070 after being subjected to one and four passes of the ECFE process, respectively. This figure also shows the normalized XRD patterns of the highest intensity peak, indicating the peak shift, which is related to long-range stress, as well as cell/sub-grain boundaries produced by the ECFE process. Furthermore, the grain size of the final pass, at the central portion of billet's cross-section, was measured by electron backscatter diffraction (EBSD) via scanning electron microscopy (SEM) with a field emission (FE) gun, which was operated at an accelerating voltage of 15 kV, beam current of 10 nA, and step size of 50 nm, using TSL OIM analysis version 5.2 (EDAX, Denizli, Turkey). Prior to the EBSD test, the surface was ground with SiC paper up to a grit size of 4000, and then electro-polished in a solution of 100 mL $HClO_4$ + 900 mL CH_3OH at $-20\,^{\circ}C$ at a voltage of 50 V.

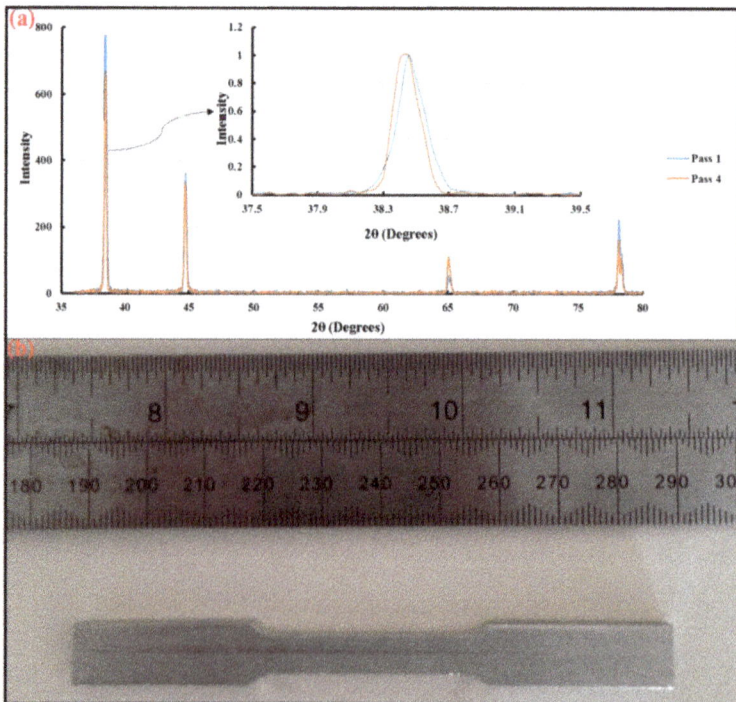

Figure 3. (a) The XRD pattern of the CP aluminum after the first and fourth passes of the ECFE process and (b) the prepared tensile testing sample of a CP aluminum billet.

3.4. Mechanical Testing

After the extrusion of Al1070, up to the four passes, various mechanical and microstructural tests were performed to determine the behavior of CP aluminum material before and after the process. A tensile test was carried out, according to the ASTM B557M, to obtain yield strength (YS), ultimate tensile strength (UTS), and elongation to failure (El%). As can be seen in Figure 3b, the gage length, gage width and sample thickness were 25 mm, 6 mm, 2 mm, respectively. In addition, Vickers micro-hardness (HV) examination was done according to ASTM E92 at the cross-section of the specimens, before and after the ECFE process, to evaluate the hardness properties. The magnitudes of imposed load and dwell time were 100 gf and 15 s, respectively. In order to prepare hardness specimens, SiC paper up to the 1000 grit and afterwards, the 3 μm diamond paste were used. HV measurements were performed ten times for each pass sample, and the average magnitude was reported. The wire-cut type of electro-discharge machining (EDM) was used to prepare the tensile test samples.

Moreover, a low velocity drop weight impact test using Instron's Dynatup 9250 HV machine (Instron, Denizli, Turkey) was employed to investigate the influence of the ECFE process on the impact behavior of CP aluminum. During this test, a ball-end dart or tup is raised to a specific height and dropped suddenly on to the test sample, as can be schematically seen in Figure 4. The magnitude of imposed impact energy was constant and equal to 40 J for all aluminum samples. The shape of the impact tup was a hemisphere with a diameter of 12.5 mm, and, in additiona, a sample with a thickness of 5 mm was circumferentially clamped using a pneumatic clamp. It was considered to be a fixed-fixed support. The time histories of the impact load and impact velocity were respectively measured and recorded via a load cell and a pair of photoelectric diodes, and the magnitudes of absorbed energy, tup velocity, and deflection at the center were derived and achieved by the use of motion equations [24,25]. These tests were performed to investigate the material grain size refinement, mechanical properties, and impact behavior of CP aluminum material during the ECFE process.

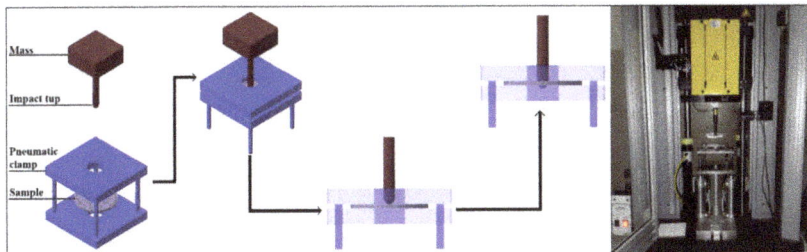

Figure 4. The schematic representation of the drop weight impact test, and, in addition, the used device.

4. Results and Discussion

4.1. Mechanical Properties

As mentioned above, the ECFE process has been successfully performed on Al1070 billets up to four passes. The engineering stress–strain curves of the ECFE aluminum, up to four passes, are presented in Figure 5, and the results including the tensile strengths, elongations to failure, and the magnitudes of hardness measurements, are listed in Table 1 and also shown in Figure 6. By considering Table 1, it can be said that enhancement of the strength values (yield strength (YS) and ultimate tensile strength (UTS)), improvement of the HV magnitudes, and reduction of the elongation percentage were achieved by utilizing this new SPD process.

Table 1. The mechanical properties and grain size of CP aluminum billets before and after the ECFE process up to four passes.

Pass Number	YS (MPa)[SD]	UTS (MPa)[SD]	El (%)[SD]	HV [SD]	Cell/sub-grain size (nm)
0	47 [1.63]	78 [0.78]	43 [0.28]	25 [1.05]	~2000
1	84 [1.34]	103 [0.78]	13.7 [0.14]	38 [1.82]	~460
2	97 [1.70]	109 [1.13]	10 [0.14]	42 [1.37]	-
3	101 [1.27]	116 [0.56]	8 [0.21]	45 [1.56]	-
4	104 [1.98]	122 [0.92]	14.7 [0.35]	46 [1.63]	~350

[SD] indicates the magnitude of standard deviation.

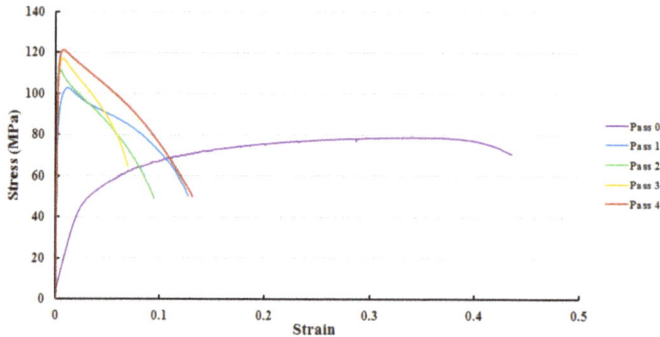

Figure 5. The engineering stress–strain curves for CP aluminum billets after the ECFE process, up to four passes.

It is obvious that, after the 1st and 4th passes of the ECFE process, 79% and 121% enhancements of the YS, 32% and 56% improvements of the UTS, and 68% and 66% reductions of the El, have been attained as compared to the annealed condition, respectively. In addition, about 52% and 84% growths of the HV magnitude were obtained after the first and fourth passes of the process, in comparison with the condition of the as-received materials, respectively. It should be pointed out that the major YS, UTS, and HV increments are achieved after the first pass, and further passes lead to improvements of these properties at a slower rate (see Figure 6). Additionally, Figure 5 shows that the improvement of the yield strength is more profound than the ultimate tensile strength. This means that the uniform plastic area, which is located between the YS and UTS points, has been limited and causes restrictions for various metal forming processes. On the other hand, an intense decrease at the elongation to failure can be seen by imposing the ECFE process. It can be noted that the reduction of the elongation to failure is high after the first pass and then it diminishes at a slower rate. The low magnitude for the elongation to failure property results in a sizeable loss of ductility after the ECFE process. Although the elongation to failure magnitude is increased slightly at the fourth pass, its magnitude is 66% lower than the aluminum under annealed conditions. The slight increase in the elongation to failure for the fourth-pass ECFE aluminum, as compared to the first pass (7%) can be related to the occurrence of grain boundary recovery (conversion of LAGBs to HAGBs) [26–28]. This can be due to a decreasing of the internal strains and accumulation of internal energy, therefore, increasing the possibility for crack formation, or it can be related to the increase of the strain rate sensitivity of the material, which causes resistance to neck formation. This improvement of the mechanical properties of material behavior using the ECFE process has also been reported for Al1050, Al6061, Al7075, and nickel during the ECAP process [26,27,29,30] and Al-3%Mg-0.2%Sc during the HPT process [31]. In addition, the ECAP process on the same material, reported by Tolaminejad and Dehghani [32], indicated that about 64% and 108% enhancement of the hardness value and approximately 202% and 267% improvement at the yield strength magnitude have been achieved after the first and fourth passes when compared to the annealed condition. The corresponding percent for the ECFE process is 52% and 84% for the HV, and 79% and 121% for the YS, respectively.

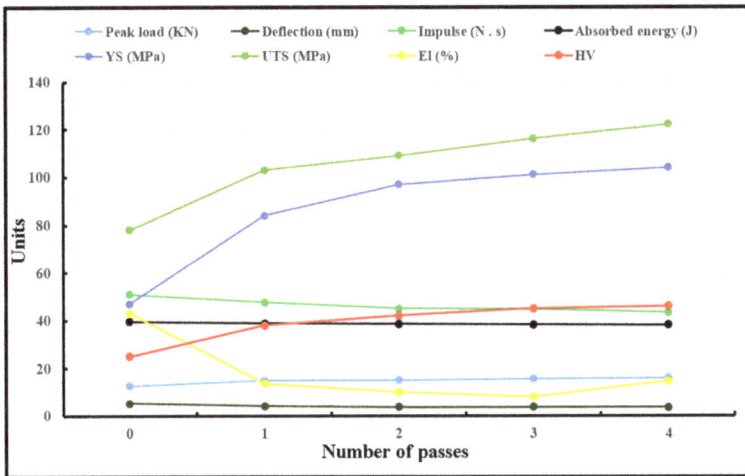

Figure 6. The variation of tensile properties, hardness behaviors, and impact characteristics of the CP aluminum, before and after the ECFE process, up to four passes.

Low velocity drop weight impact tests have been performed on the CP aluminum samples before and after the ECFE process to evaluate their impact behavior. Figure 7a,b represent the curves of load *versus* time and deflection. As seen in this figure, each curve has two regions, including loading and unloading regions. In the first part, the force slowly increases up to the maximum peak load and then it decreases sharply in the second region. The ascending region occurs due to the sample resistance to the impact force, and the descending region is related to the rebound of the tup from the sample. In other words, the required load for penetration is increased when the tup makes contact with the sample's surface. Table 2 lists the magnitudes of peak load, deflection, impulse, and absorbed energy for each pass of the ECFE sample and the trend of impact behavior *versus* pass number is also shown in Figure 6. The results indicate that the impact load increases and the deflection decreases by adding an ECFE pass number, and that the effects of the first pass ECFE process on the impact behavior is more significant than with other passes. About a 26% increment of the impact force and also 32%, 15%, and 4% decreases at the deflection, impulse, and absorbed energy, have been obtained, respectively, after the four passes of the ECFE process in comparison with the as-received condition. Thus, it can be said that the highest impact load, and also the lowest deflection, impulse, and absorbed energy can be achieved for the fourth-pass ECFE sample, which means that an aluminum sample with an enhanced strength and brittle behavior has been achieved. These results confirm the obtained tensile properties. Interestingly, the smooth slope of the curves of the annealed and first pass samples in Figure 7b indicates that the deflection is continued by the same load, which is observed in the ductile materials.

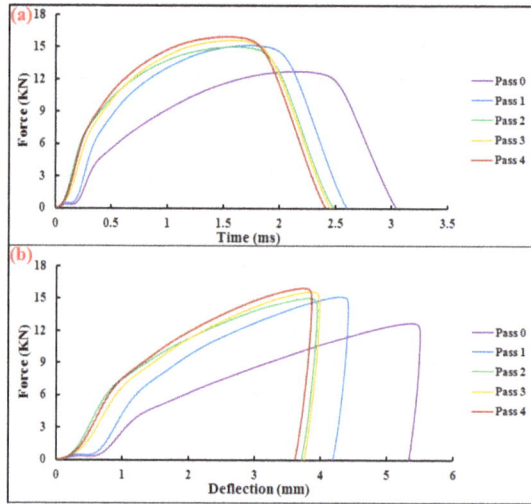

Figure 7. The (**a**) force–time and (**b**) force–deflection curves of CP aluminum billets, before and after the ECFE process, up to four passes, achieved by a drop weight impact test.

Table 2. The impact characteristics of CP aluminum billets before and after the ECFE process up to the four passes.

Pass Number	Peak load (KN)[SD]	Deflection (mm)[SD]	Impulse (N·s)[SD]	Absorbed energy (J)[SD]
0	12.72 [0.03]	5.33 [0.06]	50.99 [0.03]	39.63 [0.02]
1	15.03 [0.04]	4.19 [0.06]	47.64 [0.06]	38.92 [0.01]
2	15.17 [0.06]	3.71 [0.01]	45.03 [0.02]	38.56 [0.02]
3	15.64 [0.06]	3.75 [0.01]	44.88 [0.06]	38.31 [0.03]
4	15.98 [0.01]	3.61 [0.03]	43.30 [0.06]	38.16 [0.03]

[SD] indicates the magnitude of standard deviation.

4.2. Microstructural Characteristics

The proposed strength mechanism for the SPD materials (modified Hall–Petch relationship) is combined with the contribution of the incidental dislocation boundaries due to the statistical trapping of dislocations (LAGBs) and the geometrically necessary boundaries, because of the difference in the slip system operating with the neighboring slip systems, or local strain differences within each grain (HAGBs) [33]. This superior strengthening behavior is accompanied with a dramatic grain size reduction. Williamson-Hall analyses on the XRD patterns have been used to calculate the crystalline size of the ECFE billets, as can be seen in Figure 3a. Only three high intensity peaks of the XRD patterns of the first and final pass samples have been considered. In addition, optical microscopy (OM) has been employed to measure the average grain size of the as-received aluminum billet. Table 1 lists the average grain size of Al1070 before and after the ECFE process for the first and final passes. The results show that the ECFE process results in 77% and 82% reductions at the grain size for the 1st and 4th passes of the deformed CP Al as compared to the non-ECFE condition. Furthermore, the EBSD orientation color map reveals that the magnitude of the average grain size is about 480 nm after four passes of the ECFE process, as is represented in Figure 8. Although the calculated XRD method gives the cell or sub-grain size, and the EBSD analysis expresses the average grain size [34], it seems that the results of the EBSD image are more accurate than the theoretical approach because the XRD procedure could be affected by the strain, stress, and energy density and also other relevant parameters influencing

13

the measurement. It is clearly observed that the fraction number of LAGBs is minor and the HAGBs are the major fraction, *i.e.*, the large angle grain boundaries occupy about 87% of the microstructure. Similar results have been reported for the ECAP process of Al1070, rapid increment of the boundary misorientation angle, and the addition of HAGB fractions up to the four passes [32].

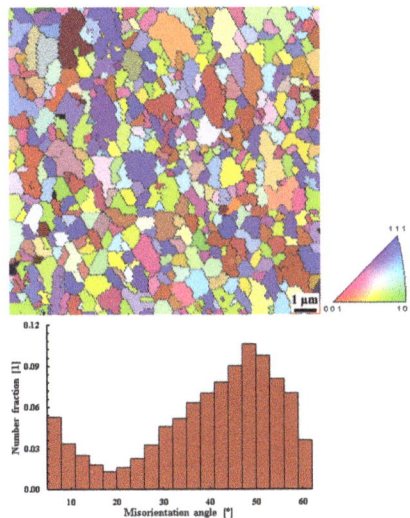

Figure 8. The orientation color map of electron backscatter diffraction for the CP aluminum billet after the four passes of the ECFE process.

5. Conclusions

In this research, the equal channel forward extrusion (ECFE) process had been proposed and introduced as a novel technique of SPD methods to fabricate UFG materials. After designing and manufacturing the die set-up, the capability of this new method has been investigated via tensile test, hardness examination, drop weight impact test, and grain size measurement on commercial pure aluminum billets, which were extruded up to four passes at room temperature. The main conclusions of this research are as follows:

- The magnitudes of yield strength, ultimate tensile strength, and Vickers micro-hardness have increased from 47 MPa, 78 MPa, and 25 HV for the annealed condition to 104 MPa, 122 MPa, and 46 HV for the fourth pass of the ECFE process, which indicate improvements of about 121%, 56%, and 84%, respectively. In addition, there is about a 66% reduction at the elongation to failure in this way. Additionally, significant enhancements in the strengthening of CP aluminum billets was achieved after the first pass of the ECFE process, which is in agreement with the hardness measurements.
- During the drop weight impact test, the magnitudes of peak load, deflection, impulse, and absorbed energy have increased from 12.72 KN, 5.33 mm, 50.99 N·s, and 39.63 J, to 15.98 KN, 3.61 mm, 43.3 N·s, and 38.16 J after four passes of the ECFE process, which means that material with the enhanced strength and brittle behavior has been attained.
- This superior improvement in the mechanical properties of the ECFE CP aluminum billet is accompanied with grain size reduction. The use of the classic Williamson–Hall method on the XRD patterns indicates that about 77% and 82% reductions have been obtained in the cell/sub-grain size of the first and fourth passes of the aluminum billets, in comparison with the annealed

condition. Additionally, the EBSD scan of the final pass indicates an average grain size of about 480 nm.

The above outcomes denote that the ECFE process can be a suitable candidate as one of the SPD methods for grain refinement and the production of UFG materials.

Author Contributions: M. Ebrahimi and F. Djavanroodi were conceived and designed this study. M. Ebrahimi wrote this manuscript and contributed in all activities. The edition was done by F. Djavanroodi. Also, experiments were performed by S.A. Nazari, H. Gholipour and C. Gode. All authors read and approved the manuscript.

Conflicts of Interest: The authors declare no conflict of interest.

References

1. Valiev, R.Z.; Langdon, T.G. Principles of equal-channel angular pressing as a processing tool for grain refinement. *Prog. Mater. Sci.* **2006**, *51*, 881–981. [CrossRef]
2. Zhilyaev, A.P.; Langdon, T.G. Using high-pressure torsion for metal processing: Fundamentals and applications. *Prog. Mater. Sci.* **2008**, *53*, 893–979. [CrossRef]
3. Kim, W.J.; An, C.W.; Kim, Y.S.; Hong, S.I. Mechanical properties and microstructures of an AZ61 Mg Alloy produced by equal channel angular pressing. *Scr. Mater.* **2002**, *47*, 39–44. [CrossRef]
4. Sergueeva, A.V.; Stolyarova, V.V.; Valiev, R.Z.; Mukherjee, A.K. Enhanced superplasticity in a Ti-6Al-4V alloy processed by severe plastic deformation. *Scr. Mater.* **2000**, *43*, 819–824. [CrossRef]
5. Rafizadeh, E.; Mani, A.; Kazeminezhad, M. The effects of intermediate and post-annealing phenomena on the mechanical properties and microstructure of constrained groove pressed copper sheet. *Mater. Sci. Eng. A* **2009**, *515*, 162–168. [CrossRef]
6. Beygelzimer, Y.; Varyukhin, V.; Synkov, S.; Orlov, D. Useful properties of twist extrusion. *Mater. Sci. Eng. A* **2009**, *503*, 14–17. [CrossRef]
7. Kwan, C.C.F.; Wang, Z. Cyclic deformation of ultra-fine grained commercial purity aluminum processed by accumulative roll-bonding. *Materials* **2013**, *6*, 3469–3481. [CrossRef]
8. Lee, J.C.; Seok, H.K.; Suh, J.Y. Microstructural evolutions of the Al strip prepared by cold rolling and continuous equal channel angular pressing. *Acta Mater.* **2002**, *50*, 4005–4019. [CrossRef]
9. Saito, Y.; Tsuji, N.; Utsunomiya, H.; Sakai, T.; Hong, R.G. Ultra-fine grained bulk aluminum produced by accumulative roll-bonding (ARB) process. *Scr. Mater.* **1998**, *39*, 1221–1227. [CrossRef]
10. Shin, D.H.; Park, J.J.; Kim, Y.S.; Park, K.T. Constrained groove pressing and its application to grain refinement of aluminum. *Mater. Sci. Eng. A* **2002**, *328*, 98–103. [CrossRef]
11. Tóth, L.S.; Arzaghi, M.; Fundenberger, J.J.; Beausir, B.; Bouaziz, O.; Arruffat-Massion, R. Severe plastic deformation of metals by high-pressure tube twisting. *Scr. Mater.* **2009**, *60*, 175–177. [CrossRef]
12. Mohebbi, M.S.; Akbarzadeh, A. Accumulative spin-bonding (ASB) as a novel SPD process for fabrication of nanostructured tubes. *Mater. Sci. Eng. A* **2010**, *528*, 180–188. [CrossRef]
13. Faraji, G.; Mashhadi, M.M.; Kim, H.S. Tubular channel angular pressing (TCAP) as a novel severe plastic deformation method for cylindrical tubes. *Mater. Lett.* **2011**, *65*, 3009–3012. [CrossRef]
14. Xue, Q.; Beyerlein, I.J.; Alexander, D.J.; Gray, G.T. Mechanisms for initial grain refinement in OFHC copper during equal channel angular pressing. *Acta Mater.* **2007**, *55*, 655–668. [CrossRef]
15. Zhao, G.; Xu, S.; Luan, Y.; Guan, Y.; Lun, N.; Ren, X. Grain refinement mechanism analysis and experimental investigation of equal channel angular pressing for producing pure aluminum ultra-fine grained materials. *Mater. Sci. Eng. A* **2006**, *437*, 281–292. [CrossRef]
16. Su, C.W.; Lu, L.; Lai, M.O. A model for the grain refinement mechanism in equal channel angular pressing of Mg alloy from microstructural studies. *Mater. Sci. Eng. A* **2006**, *434*, 227–236. [CrossRef]
17. Shin, D.H.; Kim, I.; Kim, J.; Park, K.T. Grain refinement mechanism during equal-channel angular pressing of a low-carbon steel. *Acta Mater.* **2001**, *49*, 1285–1292. [CrossRef]
18. Ebrahimi, M.; Djavanroodi, F. Experimental and numerical analyses of pure copper during ECFE process as a novel severe plastic deformation method. *Prog. Nat. Sci. Mater. Inter.* **2014**, *24*, 68–74. [CrossRef]
19. Wang, J.W.; Duan, Q.Q.; Huang, C.X.; Wu, S.D.; Zhang, Z.F. Tensile and compressive deformation behaviors of commercially pure Al processed by equal-channel angular pressing with different dies. *Mater. Sci. Eng. A* **2008**, *496*, 409–416. [CrossRef]

20. Reihanian, M.; Ebrahimi, R.; Moshksar, M.M.; Terada, D.; Tsuji, N. Microstructure quantification and correlation with flow stress of ultrafine grained commercially pure Al fabricated by equal channel angular pressing (ECAP). *Mater. Charact.* **2008**, *59*, 1312–1323. [CrossRef]

21. Mukherjee, P.; Sarkar, A.; Barat, P.; Bandyopadhyay, S.K.; Sen, P.; Chattopadhyay, S.K.; Chatterjee, P.; Chatterjee, S.K.; Mitra, M.K. Deformation characteristics of rolled zirconium alloys: A study by X-ray diffraction line profile analysis. *Acta Mater.* **2004**, *52*, 5687–5696. [CrossRef]

22. Zhang, Z.; Zhou, F.; Lavernia, E.J. On the analysis of grain size in bulk nanocrystalline materials via X-ray diffraction. *Metall. Mater. Trans. A* **2003**, *34*, 1349–1355. [CrossRef]

23. Hosseini, E.; Kazeminezhad, M. Nanostructure and mechanical properties of 0–7 strained aluminum by CGP: XRD, TEM and tensile test. *Mater. Sci. Eng. A* **2009**, *526*, 219–224. [CrossRef]

24. Múgica, J.I.; Aretxabaleta, L.; Ulacia, I.; Aurrekoetxea, J. Impact characterization of thermoformable fibre metal laminates of 2024-T3 aluminum and AZ31B-H24 magnesium based on self-reinforced polypropylene. *Compos. Part A* **2014**, *61*, 67–75. [CrossRef]

25. Liu, B.; Villavicencio, R.; Soares, C.G. On the failure criterion of aluminum and steel plates subjected to low-velocity impact by a spherical indenter. *Int. J. Mech. Sci.* **2014**, *80*, 1–15. [CrossRef]

26. Shokuhfar, A.; Nejadseyfi, O. A comparison of the effects of severe plastic deformation and heat treatment on the tensile properties and impact toughness of aluminum alloy 6061. *Mater. Sci. Eng. A* **2014**, *594*, 140–148. [CrossRef]

27. Krasilnikov, N.; Lojkowski, W.; Pakiela, Z.; Valiev, R. Tensile strength and ductility of ultra-fine-grained nickel processed by severe plastic deformation. *Mater. Sci. Eng. A* **2005**, *397*, 330–337. [CrossRef]

28. Bystrzycki, J.; Fraczkiewicz, A.; Lyszkowski, R.; Mondon, M.; Pakiela, Z. Microstructure and tensile behavior of Fe–16Al-based alloy after severe plastic deformation. *Intermetallics* **2010**, *18*, 1338–1343. [CrossRef]

29. Puertas, I.; Luis Pérez, C.J.; Salcedo, D.; León, J.; Fuertes, J.P.; Luri, R. Design and mechanical property analysis of AA1050 turbine blades manufactured by equal channel angular extrusion and isothermal forging. *Mater. Des.* **2013**, *52*, 774–784. [CrossRef]

30. Shaeri, M.H.; Salehi, M.T.; Seyyedein, S.H.; Abutalebi, M.R.; Park, J.K. Microstructure and mechanical properties of Al-7075 alloy processed by equal channel angular pressing combined with aging treatment. *Mater. Des.* **2014**, *57*, 250–257. [CrossRef]

31. Harai, Y.; Edalati, K.; Horita, Z.; Langdon, T.G. Using ring samples to evaluate the processing characteristics in high-pressure torsion. *Acta Mater.* **2009**, *57*, 1147–1153. [CrossRef]

32. Tolaminejad, B.; Dehghani, K. Microstructural characterization and mechanical properties of nanostructured AA1070 aluminum after equal channel angular extrusion. *Mater. Des.* **2012**, *34*, 285–292. [CrossRef]

33. Luo, P.; McDonald, D.T.; Xu, W.; Palanisamy, S.; Dargusch, M.S.; Xia, K. A modified Hall–Petch relationship in ultrafine-grained titanium recycled from chips by equal channel angular pressing. *Scr. Mater.* **2012**, *66*, 785–788. [CrossRef]

34. Azimi, A.; Tutunchilar, S.; Faraji, G.; Besharati Givi, M.K. Mechanical properties and microstructural evolution during multi-pass ECAR of Al1100–O alloy. *Mater. Des.* **2012**, *42*, 388–394. [CrossRef]

metals

MDPI

Article

Production of Bulk Metallic Glasses by Severe Plastic Deformation

Lisa Krämer [1,*], Karoline S. Kormout [1], Daria Setman [2], Yannick Champion [3] and Reinhard Pippan [1,4]

[1]   Erich Schmid Institute of Materials Science, Austrian Academy of Sciences, Leoben 8700, Austria; karoline.kormout@oeaw.ac.at (K.S.K.); reinhard.pippan@oeaw.ac.at (R.P.)
[2]   Physics of Nanostructured Materials, Faculty of Physics, University of Vienna, Vienna 1090, Austria; daria.setman@univie.ac.at
[3]   Institut de Chimie et des Matériaux Paris-Est, Université Paris-Est Créteil, Thiais 94320, France; champion@icmpe.cnrs.fr
[4]   Department of Materials Physics, University of Leoben, Leoben 8700, Austria
*   Author to whom correspondence should be addressed; lisa.kraemer@stud.unileoben.ac.at; Tel.: +43-3842-804-228

Academic Editor: Heinz Werner Hoppel
Received: 31 March 2015; Accepted: 23 April 2015; Published: 30 April 2015

Abstract: The aim of this study was to show the possibility to produce bulk metallic glass with severe plastic deformation. High pressure torsion was used to consolidate Zr-based metallic glass powder and deform it further to weld the powder particles together. The produced samples were investigated with Scanning electron microscope (SEM), Transmission electron microscope (TEM), Differential scanning calorimetry (DSC) and X-ray diffraction (XRD) to check if the specimens are fully dense and have an amorphous structure. The results show that the specimens remain amorphous during high pressure torsion and the density depends on the applied strain. Additional Vickers hardness measurements enable a comparison with literature and show for Zr-based metallic glass powder typical values (approximately 500 HV).

Keywords: severe plastic deformation; bulk metallic glass; Zr

1. Introduction

Metallic glasses are a new class of metals which were discovered in the 1950s [1–4]. As a consequence of their amorphous structure, they have no crystal defects as grain boundaries or dislocations [2,5,6]. This microstructural state changes the mechanism of deformation and subsequently their mechanical properties. Their high yield strength and high elastic strain qualifies them for various applications [7,8]. But as any new developed material they also harbor some obstacles. Metallic glasses are not easy to produce as crystallization of the melt must be prevented. This can be difficult depending on the used composition and the aimed large dimensions [9,10]. A solution for producing bulk metallic glasses from not so stable metallic glass compositions is to produce metallic glass powder and form the bulk metallic glass in an additional step. Severe plastic deformation (SPD) can be this additional step [11,12] as well as a powder metallurgy process [13–15]. SPD techniques were used in the past to obtain ultra-fine grained (ufg) and nanocrystalline (nc) specimens [16–19]. A wide range of materials was processed with SPD and even immiscible systems were deformed to obtain nanocomposites [20]. The flexibility of these routes is also shown in the used starting materials. Those can range from solid to powder and so the process of alloying is simplified. If High Pressure Torsion (HPT) is used, powders can simply be mixed and the achieved mixture consolidated. With the applied strain during the following HPT step the material becomes homogenous. SPD was already used to deform bulk metallic glass samples and even metal-metallic glass composites were obtained [21–25].

The aim of this study was to produce bulk metallic glass samples with SPD. The Zr-based metallic glass powder was consolidated and deformed with HPT. SEM, TEM, XRD, DSC and hardness measurements were used to investigate the specimen.

2. Experimental Section

For the production of bulk metallic glass samples a metallic glass powder with the composition $Zr_{57}Cu_{20}Al_{10}Ni_8Ti_5$ has been used. This powder was prepared by high pressure gas atomization (atomization performed by Lucas Dembinski, PERSEE, IRTES-LERMPS, Belfort France) and shows an average particle size of 25 µm [21]. SEM images of the bulk metallic glass powder can be seen in Figure 1. A contamination particle is indicated with an arrow (see Figure 1b). Energy dispersive X-ray spectroscopy (EDX) measurements (Oxford Instruments plc, Abdingdon, UK) have shown that contamination contains Mn, Si, Fe, Al and Mg. For the HPT process (Schenck, Germany; adapted hydraulic testing machine), powder was filled between the gap of the grooved HPT anvils, where it was then compacted and processed by torsion under a pressure of 8 GPa at room temperature up to 63 revolutions. The final dimensions of the samples used in this study are 6 mm in diameter and 0.6 mm in height. Some contaminating particles were too hard to be deformed simultaneously with the metallic glass powder. They, however, do not seem to influence the results of the general microstructure (their total amount is below 1%); they may affect the ductility which is not in the focus of this paper. The deformed HPT disks were cut into halves, ground and polished for SEM investigations and hardness measurements. The Vickers hardness was measured along the diameter in the cross-section of the HPT sample with a load of 0.5 kg. XRD phase analysis of the samples was performed using a 5-circle X-ray diffractometer (SmartLab, Rigaku Co., Tokyo, Japan) equipped with a source for Cu-Kα radiation. DSC was performed using a Netzsch DSC 204 (NETZSCH GmbH & Co.Holding KG, Selb/Bavaria, Germany). The heating range was between 25 and 590 °C. The heating rate was 10 K min^{-1}. For the DSC measurements undeformed powder and HPT samples, which were cut in half, were used. TEM samples were prepared by a standard procedure: grinding, polishing and dimple grinding with subsequent ion milling. TEM micrographs were recorded in top view at a radius of 2 mm (±0.3 mm due to varying dimension and position of the hole). Microstructural investigations were conducted using a SEM LEO1525 (Carl Zeiss AG, Oberkochen, Germany) and a (scanning) transmission electron microscope (S)TEM JEOL JEM 2100F (JEOL Ltd., Akishima, Japan) equipped with a C$_S$-corrector (CEOS GmbH, Heidelberg, Germany).

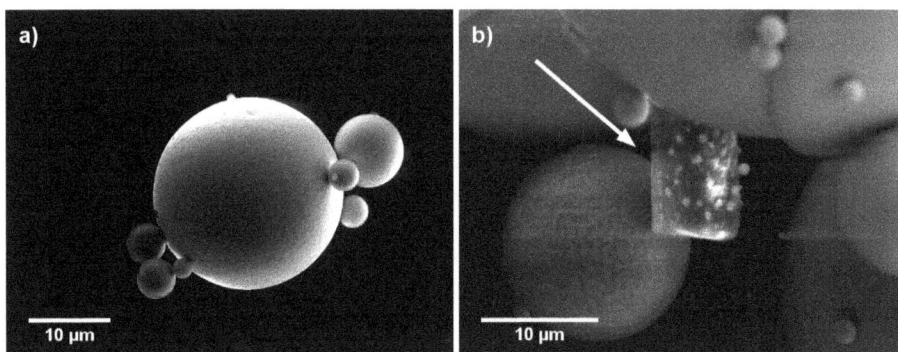

Figure 1. SEM images of the Zr-based metallic glass powder. The average particle size is 25 µm. The spherical shape (**a**) is a result of the production by high pressure gas atomization. Contaminating particle indicated with an arrow in (**b**).

3. Results and Discussion

This work studies the possibility to produce bulk metallic glass specimens by using HPT; starting with compaction of metallic glass powder and subsequent deformation. Therefore, three essential questions must be answered: Is the final sample fully dense or does the sample contain pores or cracks? Does the structure of the metallic glass change during deformation? Is the hardness of the HPT produced metallic glass comparable to values of bulk metallic glass in literature?

To answer the first question samples were investigated in the SEM to find (even very small) inhomogeneities. In HPT the torsion of the sample leads to a strain gradient along the radius of the sample. At the center the applied shear strain is nearly zero, but at the edge it increases linearly with the radius. Hence, only few HPT disks are sufficient to investigate the influence of the applied strain. The investigated sample was deformed for 30 rotations at room temperature and cut into halves. Figures 2 and 3 show the center and the edge in the cross-section of the HPT disk with a shear strain of approximately 0 and 900, respectively.

In Figure 2, crack-like defects can be seen. These cracks are not induced by deformation, they are a consequence of the insufficient deformation of the initial powder particles. The distance between these defects corresponds to the particle diameter; it is between 2 and 25 μm (see Figure 1). Those cracks can be found up to a radius of approximately 1.5 mm, which equals to a shear strain of about 600. Apparently, the deformation near the center was not large enough to enforce a full consolidation of the powder and fully weld the particles together.

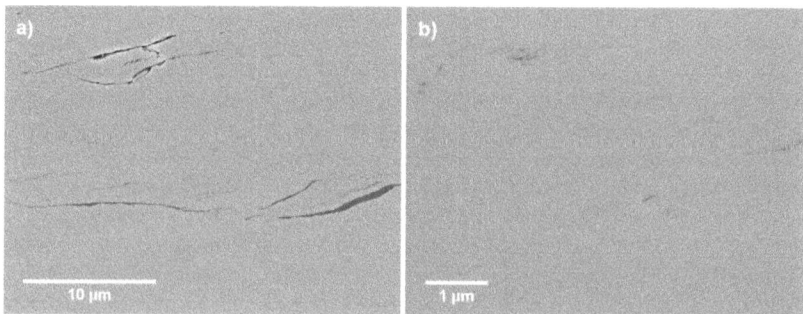

Figure 2. SEM images taken from regions near the center of the High Pressure Torsion (HPT) disk at different magnifications (lower magnification (**a**) and higher magnification (**b**)). The crack-like defects are the result of unfinished welding of the metallic glass powder. The distance between the cracks correlates with the particle size of the powder.

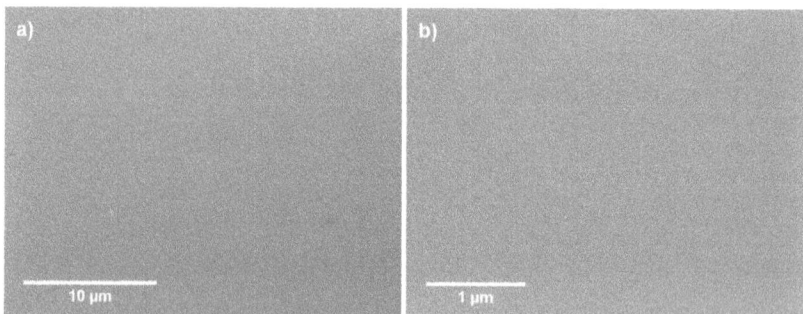

Figure 3. SEM images near the edge of the HPT disk. At low magnifications (**a**) and even at high magnifications (**b**), no cracks, pores or other inhomogeneities can be detected.

At the edge of the same specimen (Figure 3), no cracks or pores can be detected. So, these SEM images confirm that full density is reached after sufficient deformation of the specimen. No other features are visible in the SEM images due to the absence of cracks, pores and boundaries.

XRD and DSC measurements as well as TEM imaging were used to investigate the microstructure of the HPT deformed specimen. In XRD broad peaks are expected for an amorphous material, while crystalline materials would show several sharp peaks depending on the crystal structure and the lattice parameters. The position of the metallic glass peak depends on the present short-range order, which is strongly influenced by the composition. In Figure 4, the XRD results of the undeformed powder and a deformed specimen are presented. A broad peak typical for amorphous material can be seen for the HPT deformed sample (black line) and the undeformed powder (gray line). The position and width of the peaks are identical and this confirms that bulk metallic glass can be produced by HPT using metallic glass powder. For the used Zr-based metallic glass the broad peaks are in the same angle range as the peaks of crystalline Zr. Shifting of this angles is due to the other elements, such as Cu and Ni.

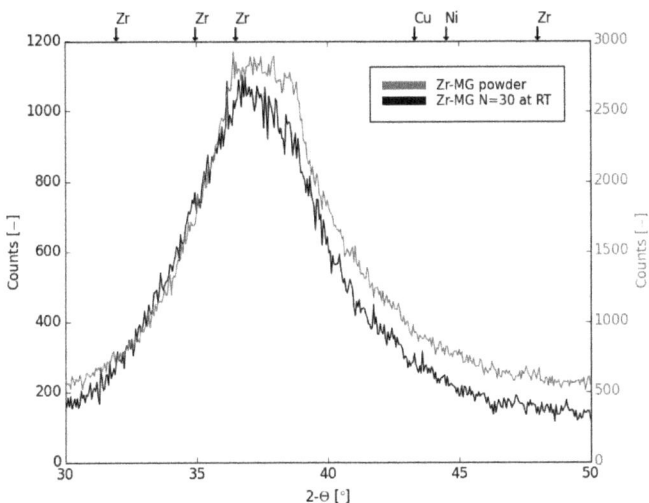

Figure 4. XRD measurement of the HPT deformed sample and the undeformed powder (gray line). The peak position and width are in both cases the same, which indicates an unchanged amorphous structure after HPT deformation. The arrows indicate the peak positions of crystalline Zr, Ni and Cu. The number of counts differs as different slits were used during the XRD measurements.

Additionally, the specimen was examined in TEM to confirm that the material is still amorphous after HPT deformation. In Figure 5, a high resolution TEM image and the diffraction pattern for the same position are shown. In the TEM image no ordered region is visible. The diffraction pattern consists of two broad rings only, which indicates-analog to the XRD measurements-an amorphous state of the sample.

DSC measurements substantiate the results of XRD and TEM. Two HPT disks with 10 and 30 turns at room temperature and undeformed powder were measured and the results can be seen in Figure 6. All three specimens crystallize at nearly the same temperature (460 °C, 457 °C and 455 °C) and the crystallization peaks have got a very similar breadth. Therefore, it can be concluded that during HPT deforming no crystallization occurs. The onset of crystallization differs for the deformed specimen and the undeformed powder. This is caused by structural relaxation during the deformation as the high compressive loading of the HPT favors a structural change to a state with higher density (which equals a more relaxed state) [26]. The relaxation in the HPT specimen causes also a shift of the glass transition temperature to higher temperatures and a stronger overlap with the crystallization peak.

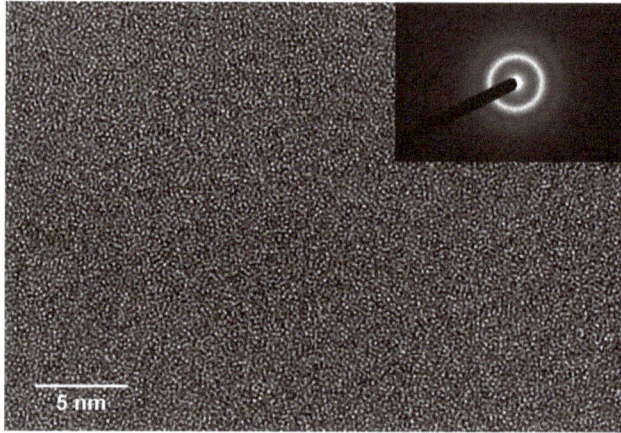

Figure 5. In the high resolution TEM image no ordered regions can be detected. The inset shows a diffraction pattern at the same position but from a much larger region (selected area diffraction (SAD) aperture size: 120 μm). The broad rings confirm the amorphous state of the specimen.

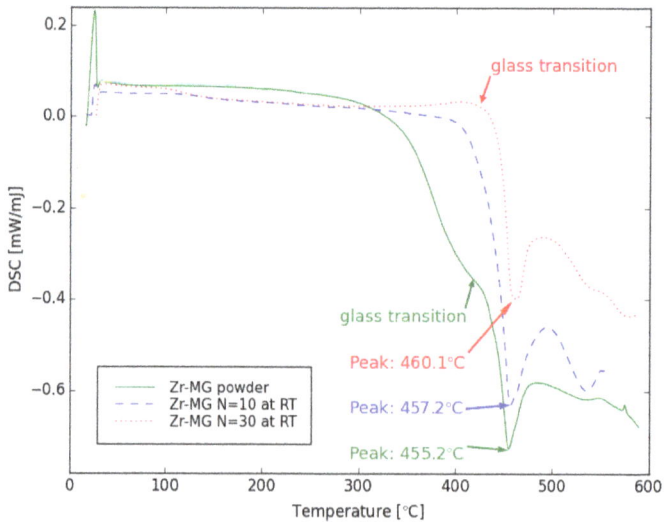

Figure 6. Differential scanning calorimetry (DSC) measurements exclude a crystallization during HPT deformation. The powder as well as the HPT disks show a distinctive crystallization peak at 457 °C, 455 °C and 460 °C, respectively. The peak breadth and position are nearly the same, which indicates an amorphous structure for all. The difference in the onset of crystallization and in the transition temperature is caused by structural relaxation due to the HPT process.

The Vickers hardness was measured to get information about the mechanical properties of the processed bulk metallic glass. In Figure 7, the results are shown for three specimens, which were deformed at room temperature for three different numbers of rotations (10, 30 and 63 turns). After ten turns the specimen contains crack-like defects all over the diameter as the applied strain was not high enough to fully consolidate the powder. This leads to a decrease of hardness. Only near the edge, the hardness nearly approaches the value of the other samples. The hardness of the specimens with 30 and 63 turns is similar. Near the center a minimum hardness with about 450 HV is measured and at the edge it increases up to 500 HV. From the fact that the specimens with 30 and 63 rotation show nearly the same hardness it can be concluded that the hardness will not increase further more, even when additional deformation is applied. Hence, it is obvious that a saturation state is reached at the edge of the samples. The values of hardness for Zr-based glass in literature are in the range of 420 —-470 HV (Stolpe *et al.* [24]), 455 —-475 HV (Luo *et al.* [27]), 480 HV (Champion and Perrière [28]) and 500 HV (Chen [29] and Xu *et al.* [30]). The variation in hardness seems to be caused by different compositions and production routes. Furthermore, one should take into account that HPT generated bulk metallic glass is heavily deformed.

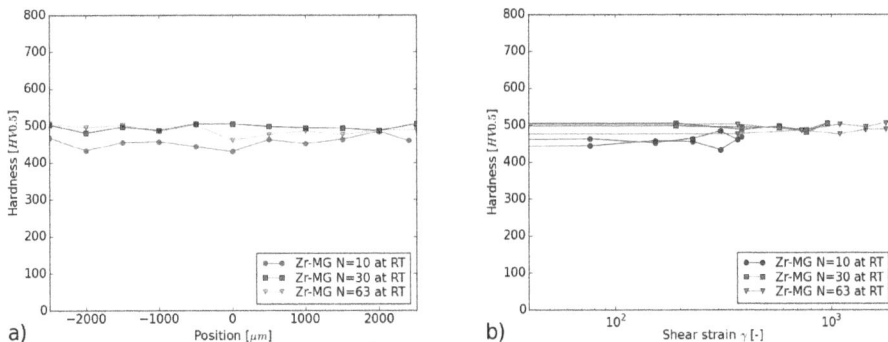

Figure 7. The results of Vickers hardness measurements can be seen. The hardness depending on the radius (**a**) of the HPT disk is shown as well as the hardness depending on the shear strain (**b**). Specimens with a higher number of applied rotations, show higher shear strain. For the sample with 10 turns the lowest hardness was measured. This is a result of incomplete welding of the powder and therefore crack-like defects. The specimens with 30 and 63 rotations show a similar hardness, which indicates that an equilibrium state is reached.

The SEM micrographs indicate that HPT is a process capable to consolidate metallic glass powder by welding the particles together. In the past HPT has been already used to produce nanocomposites by co-deforming of two different materials. Hence, the next step is producing metallic glass composites. The second component next to a metallic glass can be a crystalline metal, but also another metallic glass. Mixing crystalline and amorphous metals together changes the deformation behavior and so, the formation of one fatal shear band can be prevented and the mechanical properties could be improved [21,31,32]. Metallic glass/metallic glass composites will also show different properties, because the two metallic glasses differ in the chemical composition and the near range ordering will vary. The interfaces between the two metallic glasses will strongly influence the properties of the composite.

4. Conclusions

(i) Bulk metallic glass specimens can be produced by consolidation and deforming metallic glass powder with HPT; (ii) fully dense samples without cracks and pores can be achieved as long as sufficient strain ($\gamma = 600$) is applied to weld the powder particles; the necessary strain is significantly larger than in the case of consolidation of metal powder, due to the more localized deformation in metallic glasses; (iii) during HPT at room temperature no crystallization of the Zr-based metallic glass was observed, the specimens were found to be fully amorphous; (iv) a hardness of up to 500 HV is measured.

Metals 2015, 5, 720–729

With this processing route it is possible to create new materials such as metal/metallic glass composites and metallic glass/metallic glass composites (two different metallic glasses). Those materials are expected to show interesting properties as the new interfaces will change dramatically the properties of metallic glasses.

Acknowledgments: Funding for this work has been provided by the European Research Council under ERC Grant Agreement No. 340185 USMS and by the Austrian Science Fund (FWF) under Project: T512-20.

Author Contributions: Lisa Krämer processed the samples, did the SEM and XRD investigations, carried out the hardness measurements and created the initial draft. Karoline Kormout carried out high resolution TEM imaging and recorded diffraction patterns. Daria Setman characterized the material with DSC measurements. Yannick Champion provided the metallic powder. Reinhard Pippan formulated the idea of this work. All authors discussed the results, contributed ideas, participated in the manuscript preparation and approved the final manuscript.

Conflicts of Interest: The authors declare no conflict of interest.

References

1. Greer, A.L. Metallic glasses ... on the threshold. *Mater. Today* **2009**, *12*, 14–22. [CrossRef]
2. Löffler, J. Bulk metallic glasses. *Intermetallics* **2003**, *11*, 529–540. [CrossRef]
3. Ashby, M.F.; Greer, A.L. Metallic glasses as structural materials. *Scr. Mater.* **2006**, *54*, 321–326. [CrossRef]
4. Wang, W.H.; Dong, C.; Shek, C.H. Bulk metallic glasses. *Mater. Sci. Eng. R* **2004**, *44*, 45–89. [CrossRef]
5. Packard, C.E.; Schuh, C.A. Initiation of shear bands near a stress concentration in metallic glass. *Acta Mater.* **2007**, *55*, 5348–5358. [CrossRef]
6. Liu, X.J.; Xu, Y.; Hui, X.; Lu, Z.P.; Li, F.; Chen, G.L.; Lu, J.; Liu, C.T. Metallic liquids and glasses: Atomic order and global packing. *Phys. Rev. Lett.* **2010**, *105*, 155501. [CrossRef] [PubMed]
7. Schroers, J.; Kumar, G.; Hodges, T.M.; Chan, S.; Kyriakides, T.R. Bulk metallic glasses for biomedical applications. *JOM* **2009**, *61*, 21–29. [CrossRef]
8. Inoue, A.; Wang, X.M.; Zhang, W. Developments and applications of bulk metallic glasses. *Rev. Adv. Mater. Sci.* **2008**, *18*, 1–9.
9. Lu, Z.P.; Liu, C.T. A new glass-forming ability criterion for bulk metallic glasses. *Acta Mater.* **2002**, *50*, 3501–3512. [CrossRef]
10. Li, Y.; Poon, S.; Shiflet, G.; Xu, J.; Kim, D.; Löffler, J. Formation of bulk metallic glasses and their composites. *MRS Bull.* **2007**, *32*, 624–628. [CrossRef]
11. Azushima, A.; Kopp, R.; Korhonen, A.; Yang, D.Y.; Micari, F.; Lahoti, G.D.; Groche, P.; Yanagimoto, J.; Tsuji, N.; Rosochowski, A.; *et al.* Severe plastic deformation (SPD) processes for metals. *CIRP Ann.-Manuf. Technol.* **2008**, *57*, 716–735. [CrossRef]
12. Krämer, L.; Wurster, S.; Pippan, R. Deformation behavior of Cu-composites processed by HPT. *IOP Conf. Ser.: Mater. Sci. Eng.* **2014**, *63*, 1–9. [CrossRef]
13. Nowak, S.; Perriére, L.; Dembinski, L.; Tusseau-Nenez, S.; Champion, Y. Approach of the spark plasma sintering mechanism in $Zr_{57}Cu_{20}Al_{10}Ni_8Ti_5$ metallic glass. *J. Alloys Comp.* **2011**, *509*, 1011–1019. [CrossRef]
14. Xie, G.; Louzguine-Luzgin, D.; Kimura, H.; Inoue, A. Ceramic Particulate Reinforced $Zr_{55}Cu_{30}Al_{10}Ni_5$ Metallic Glassy Matrix Composite Fabricated by Spark Plasma Sintering. *Mater. Trans.* **2007**, *48*, 1600–1604. [CrossRef]
15. Xie, G.; Louzguine-Luzgin, D.V.; Kimura, H.; Inoue, A. Nearly full density $Ni_{52.5}Nb_{10}Zr_{15}Ti_{15}Pt_{7.5}$ bulk metallic glass obtained by spark plasma sintering of gas atomized powders. *Appl. Phys. Lett.* **2007**, *90*, 241902. [CrossRef]
16. Valiev, R.Z.; Islamgaliev, R.K.; Alexandrov, I.V. Bulk nanostructured materials from severe plastic deformation. *Progr. Mater. Sci.* **2000**, *2*, 103–189. [CrossRef]
17. Valiev, R.Z.; Estrin, Y.; Horita, Z.; Langdon, T.G.; Zechetbauer, M.J.; Zhu, Y.T. Producing bulk ultrafine-grained materials by severe plastic deformation. *JOM* **2006**, *58*, 33–39. [CrossRef]
18. Estrin, Y.; Vinogradov, A. Extreme grain refinement by severe plastic deformation: A wealth of challenging science. *Acta Mater.* **2013**, *61*, 782–817. [CrossRef]
19. Sabirov, I.; Pippan, R. Formation of a W25%Cu nanocomposite during high pressure torsion. *Scr. Mater.* **2005**, *52*, 1293–1298. [CrossRef]

20. Bachmaier, A.; Pippan, R. Generation of metallic nanocomposites by severe plastic deformation. *Int. Mater. Rev.* **2013**, *58*, 41–62. [CrossRef]

21. Sauvage, X.; Champion, Y.; Pippan, R.; Cuvilly, F.; Perriere, L.; Akhatova, A.; Renk, O. Structure and properties of a nanoscaled composition modulated metallic glass. *J. Mater. Sci.* **2014**, *49*, 5640–5645. [CrossRef]

22. Sort, J.; Ile, D.C.; Zhilyaev, A.P.; Concustell, A.; Czeppe, T.; Stoica, M.; Surinach, S.; Eckert, J.; Baro, M.D. Cold-consolidation of ball-milled Fe-based amorphous ribbons by high pressure torsion. *Scr. Mater.* **2004**, *50*, 1221–1225. [CrossRef]

23. Van Steenberge, N.; Hobor, S.; Surinach, S.; Zhilyaev, A.; Houdellier, F.; Mompiou, F.; Baro, M.D.; Revesz, A.; Sort, J. Effects of severe plastic deformation on the structure and thermo-mechanical properties of $Zr_{55}Cu_{30}Al_{10}Ni_5$ bulk metallic glass. *J. Alloys Comp.* **2010**, *500*, 61–67. [CrossRef]

24. Stolpe, M.; Kruzic, J.; Busch, R. Evolution of shear bands, free volume and hardness during cold rolling of a Zr-based bulk metallic glass. *Acta Mater.* **2014**, *64*, 231–240. [CrossRef]

25. Shahabi, H.S.; Scudino, S.; Kühn, U.; Eckert, J. Metallic glass-steel composite with improved compressive plasticity. *Mater. Des.* **2014**, *59*, 241–245. [CrossRef]

26. Concustell, A.; Baro, M.D.; Mear, F.O.; Surinach, S.; Greer, A.L. Structural relaxation and rejuvenation in a metallic glass induced by shot-peening. *Philos. Mag. Lett.* **2009**, *89*, 831–840. [CrossRef]

27. Luo, K.; Li, W.; Zhang, H.Y.; Su, H.L. Changes of hardness and electronic work function of $Zr_{41.2}Ti_{13.8}Cu_{12.5}Ni_{10}Be_{22.5}$ bulk metallic glass on annealing. *Philos. Mag. Lett.* **2011**, *91*, 237–245. [CrossRef]

28. Champion, Y.; Perriére, L. Strain Gradient in Micro-Hardness Testing and Structural Relaxation in Metallic Glasses. *Adv. Eng. Mater.* 2014. [CrossRef]

29. Chen, X. Structure and hardness evolution of the scale of a Zr-based metallic glass during oxidation. *J. Non-Cryst. Solids* **2013**, *362*, 140–146. [CrossRef]

30. Xu, Y.; Zhang, Y.; Li, J.; Hahn, H. Enhanced thermal stability and hardness of $Zr_{46}Cu_{39.2}Ag_{7.8}Al_7$ bulk metallic glass with Fe addition. *Mater. Sci. Eng. A* **2010**, *527*, 1444–1447. [CrossRef]

31. Inoue, A.; Zhang, T.; Kim, Y.H. Synthesis of high strength bulk amorphous Zr-Al-Ni-Cu-Ag alloys with a nanoscale secondary phase. *Mater. Trans. JIM* **1997**, *38*, 749–755. [CrossRef]

32. Hofmann, D.C. Bulk metallic glasses and their composites: A brief history of diverging fields. *J. Mater.* **2013**, *2013*, 1–8. [CrossRef]

Chapter 2:
UFG Titanium Alloys

Article

Nanocrystalline Ti Produced by Cryomilling and Consolidation by Severe Plastic Deformation

Irina Semenova [1,*], Ilana Timokhina [2], Rinat Islamgaliev [1], Enrique Lavernia [3] and Ruslan Valiev [4]

[1] Institute of Physics of Advanced Materials, Ufa State Aviation Technical University, 12 K. Marx str., Ufa 450000, Russia; saturn@mail.rb.ru

[2] Institute for Frontier Material, Geelong Technology Precinct, Deakin University, 75 Pigdons Road, Waurn Ponds, Victoria 3216, Australia; ilana.timokhina@monash.edu

[3] Department of Chemical Engineering and Materials Science, University of California, Davis, CA 95616, USA; lavernia@ucdavis.edu

[4] Laboratory for Mechanics of Bulk Nanostructured Materials, Saint Petersburg State University, Universitetsky pr. 28, Peterhof 198504, Saint Petersburg, Russia; rzvaliev@mail.rb.ru

* Author to whom correspondence should be addressed; semenova-ip@mail.ru; Tel./Fax: +7-347-2733422.

Academic Editor: Heinz Werner Höppel

Received: 18 December 2014; Accepted: 30 January 2015; Published: 5 February 2015

Abstract: We report on a study of the nanocrystalline structure in Ti, which was produced by cryogenic milling followed by subsequent consolidation via severe plastic deformation using high pressure torsion. The mechanisms that are believed to be responsible for the formation of grains smaller than 40 nm are discussed and the influence of structural characteristics, such as nanometric grains and oxide nanoparticles, on Ti hardening is established.

Keywords: Ti; cryomilling; high pressure torsion; nanostructure; strengthening

1. Introduction

For the past decade or so, the interest in the application of severe plastic deformation (SPD) techniques to manufacture nanostructured Ti has been motivated by several factors [1–15]. First, the reports on the formation of an ultrafine-grained structure from SPD processing and its influence on the mechanical behavior have led to interesting fundamental questions [1–13]. Second, the published studies suggest that nanostructuring of Ti and its alloys can result in significant enhancements in strength, ductility and fatigue life [10,14], which provide a pathway to the structural and medical applications.

Grain refinement in commercially pure Ti by various SPD methods typically leads to ultrafine grain (UFG) structures with an average grain size in the range of 100–600 nm [3–6,12–15]. However, it has been reported that the consolidation of the metal powders via SPD methods can be implemented to prepare high-density samples with grain sizes that are smaller than 50 nm [16,17]. In the view of the above discussion, in the current study, we report on a novel processing approach for producing nanocrystalline commercially pure (CP) Ti with grain sizes that are smaller than 40 nm. Accordingly, we describe experimental results obtained by consolidating cryomilled CP Ti using the SPD techniques and discuss the influence of the nanostructure on the mechanical behavior.

2. Results and Discussion

Figure 1 shows a SEM image of the Ti powder after pre-compaction (a) and after subsequent consolidation by high pressure torsion (HPT) (b). The separated powder particles with irregular morphologies and an average particle size of about 17 μm are evident (see

Figure 1a). The microstructural analysis revealed the presence of pores between the powder particles (see Figure 1a); the volume fraction of porosity was estimated to be approximately 40% of the total area. The volume fraction of porosity decreased significantly after HPT (see Figure 1b), with separate pores smaller than 1 μm and a fraction <3 vol.% in the central part of the sample (within a radius of up to 2 mm from the center). The pores were not observed in the peripheral region of the disk.

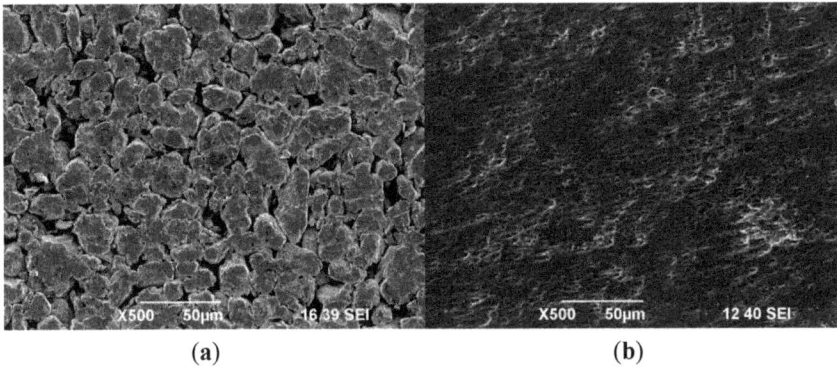

(a)   (b)

Figure 1. (a) The morphology of Ti Grade 2 powder produced by cryogenic milling and pre-compacted under a pressure of 1.95 GPa; (b) the sample's surface in the peripheral part of the sample after five turns of high pressure torsion (HPT) at 573 K.

Figure 2 shows bright-field and dark-field TEM images of the microstructure at the peripheral zone (approximately 3 mm from the center) of the sample after HPT. As noted above, this region does not show any significant porosity. The microstructure of Ti consists of ultrafine grains with an average size ranging from 25 to 40 nm (see Figure 2a). However, some coarser grains with an average size of 120 ± 50 nm were also observed in the microstructure (see Figure 2b). The fraction of the large grains in the microstructure did not exceed 7%. It is likely that the large grains in the Ti sample were formed during HPT through coalescence of the initial small grains in the cryomilled Ti powders. A similar phenomenon was observed in Cu powders subjected to HPT [18]. In this study a possible mechanism of non-uniform grain growth can be also associated with dynamic recrystallization, because SPD processing was performed at elevated deformation temperatures ($T = 573$ K) when grain boundary diffusional processes are activated.

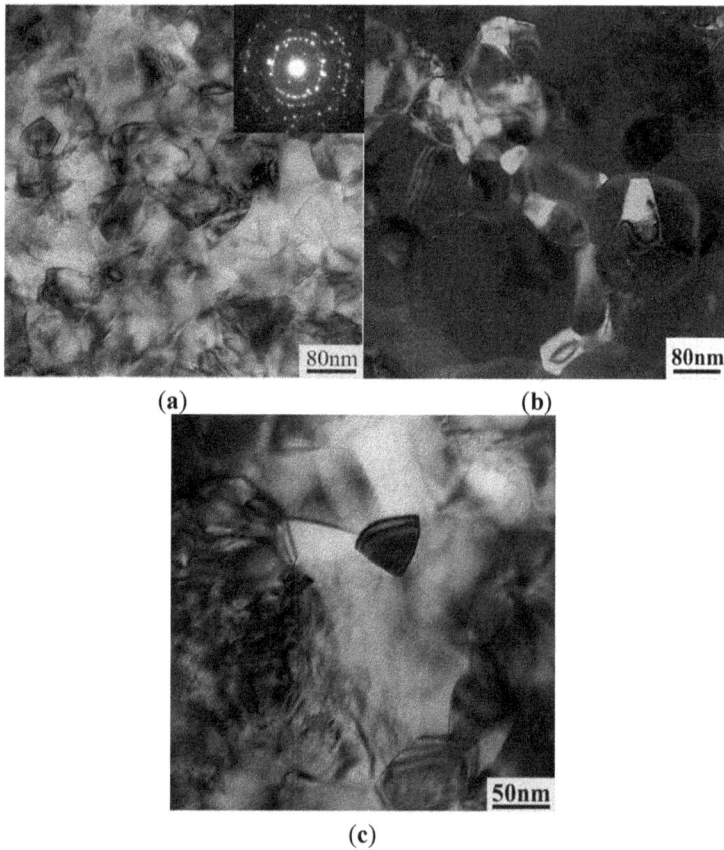

Figure 2. (a,b) Bright-field and dark-field images of ultrafine grains formed in peripheral part of the sample; (c) ultrafine grains.

The appearance of non-uniform contrast inside the grains in the dark-field images indicates the presence of a high lattice distortion (see Figure 2c). The total dislocation density at grain interiors was estimated to be approximately 2×10^{14} m^{-2}. The spots on the SAED patterns formed concentric circles (see Figure 2). This feature is consistent with the formation of high-angle boundaries, which is typical of nanostructured materials produced by HPT under conditions that involve a high number of rotations [1].

The microstructure in the central region of the sample revealed a non-uniform distribution of grain sizes ranging from 25 to 240 nm (see Figure 3). Such non-uniformity of the structure formed in the center of HPT samples can be attributed to the competing processes of deformation, dynamic recrystallization and grain growth, which are likely to be activated at elevated temperatures (573 K) [13]. The microhardness measurements along the disk radius from a sample center to its periphery showed a decrease in hardness from 4050 MPa in the peripheral region to 3600 MPa in the sample center. However, the microhardness was significantly higher in comparison with that of nanostructured monolithic Ti after HPT at room temperature reported in [4], where the minimum grain size was about 80 nm, and the microhardness averaged 3000 MPa.

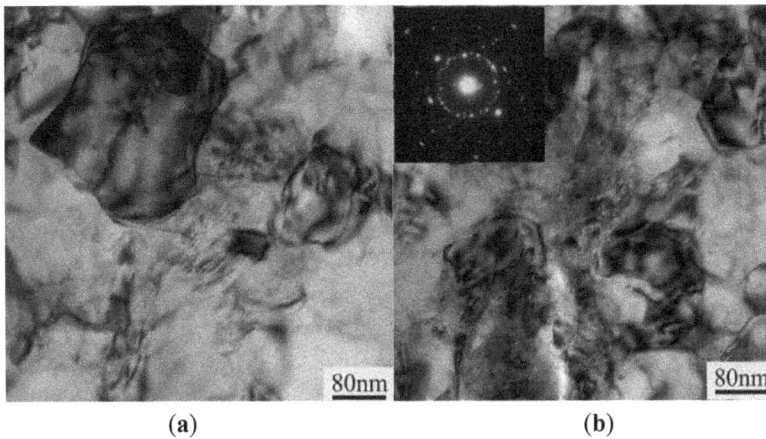

(a)   (b)

Figure 3. (a) TEM micrographs coarse grains; (b) ultrafine grains in the center of the sample.

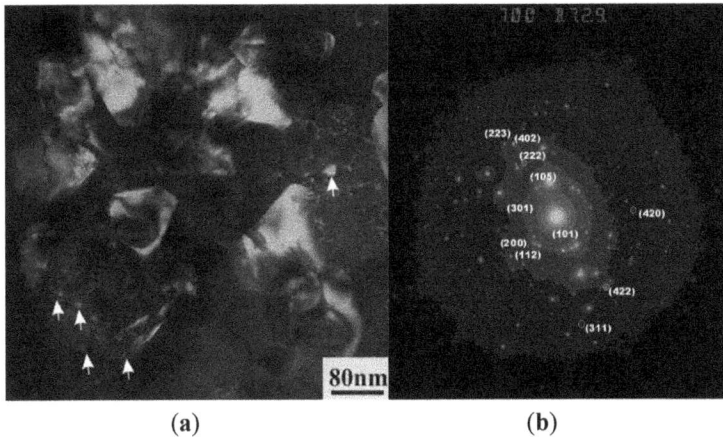

(a)   (b)

Figure 4. (a) TEM dark-field images and (b) diffraction pattern showing precipitates.

As is known, the strong affinity between Ti and O leads to the formation of a thin oxide layer in Ti, regardless of the environmental conditions used during ball milling [18,19]. Accordingly, during structural analysis of the HPT samples, particular attention was paid to the dispersed nanoparticles with a size less than 10 nm, which we observed along the grain boundaries (see Figure 4). Diffraction patterns taken from the area of 0.5 µm showed the spots with interplanar distances that correspond not only to Ti, but also to its oxides and nitrides, such as TiO_2 and TiN (see Figure 4b). Analysis of the TEM-images of the structure showed that the volume fraction of such oxide particles was very small and did not exceed 0.1% by volume.

The results presented herein support the hypothesis that it is possible to synthesize nanostructured CP Ti Grade 2 samples of high density by using a new approach involving HPT processing of powder after cryomilling. In this work, we have managed to produce a nanocrystalline structure with a mean grain size below 40 nm in CP Ti for the first time. It has been demonstrated in an earlier study [17] that the obtained nanostructure depends greatly on the size of the initial powder particles and the method of their preparation. It is known that, during milling, the material is subjected to high-strain-rate deformation of a very high degree. Here, a high level of internal stresses is created

30

due to a high density of dislocations, declinations, vacancies and other defects of the crystal lattice, introduced during deformation [19]. As the deformation degree increases, sub-boundaries forming in the powder particles transform into new high-angle boundaries, and the efficiency of structure refinement grows considerably in the conditions of low temperatures as a result of cyclic plastic deformation during milling in combination with limited recovery at low temperatures [19]. One benefit of cryogenic milling in an inert medium (in liquid nitrogen or argon) is that it hinders recovery processes (by virtue of the low mobility of defects at low temperatures) and thereby diminishes the time required to attain a nanocrystalline microstructure, as compared to the mechanical alloying, for example [19]. After cryogenic milling the average size of nanocrystals forming in the particles can reach 20–30 nm [19].

Subsequent compaction of ultradispered powders should, first, ensure the fullest consolidation of the sample, and second, preserve the nanostructure of the initial ultradispersed powder [20,21]. To attain a high density of samples, sintering and/or pressing at elevated temperatures is normally used during compaction, but an intensive recrystallization occurring in the material during this process leads to a noticeable grain coarsening [21]. It is evident that the preservation of the nanocrystalline range of grain sizes can be achieved only through a decrease in the temperature and/or an increase in the pressure during pressing. As is shown in the results of present work, an efficient method of consolidation is severe plastic deformation via high pressure torsion (6 GPa) at a temperature of 573 K, at which recrystallization processes in Ti are impeded. Nevertheless, note should be made that we did not manage to completely exclude recrystallization processes in Ti at the given temperature due to high internal stresses accumulated in the particles after their milling. As a result of such intensive straining, macro-and micro-pores were practically absent in the microstructure of the samples, and the minimum grain size reached 40 nm. At the same time, the non-uniformity of deformation in the central and peripheral parts of the sample led to a noticeable inhomogeneity of the microstructure, which affected the microhardness values (3600 and 4050 MPa, respectively).

Figure 5 shows experimental points that are plotted for CP Ti Grade 2 processed by ECAP with a grain size of 600 nm [15], processed by HPT at room temperature when the average grain size reached 80 nm [4], and for Ti Grade 2 with a nanocrystalline structure and an average grain size of around 40 nm produced within the present work ($H\mu \sim 3\sigma_{0.2}$). For the sake of comparison, the Hall-Petch relationship (1) was extrapolated to the ultrafine-grained region, where the grain size dependence of yield stress in Ti is described by the following equation:

$$\sigma = \sigma_0 + K_{H\text{-}P}d^{-1/2} \tag{1}$$

where σ_0 is the lattice friction stress, and $K_{H\text{-}P}$ is the hardening coefficient in the Hall-Petch equation. The variation σ_0 and $K_{H\text{-}P}$ can be due to the difference in material composition and process. It should be noted that these σ_0 and $K_{H\text{-}P}$ values are somewhat different to those reported previously, for example, well-annealed pure Ti had $\sigma_0 = 78$ MPa and $K_{H\text{-}P} = 0.40$ MPa·m$^{1/2}$ [22], CP Ti subjected to cryogenic channel die compression with an intercept of 249 MPa, and a slope of 0.27 MPa·m$^{1/2}$ [23]. In this work, the value of σ_0 was experimentally determined as the yield stress (σ_{02}) of commercially pure Ti Grade 2 with a grain size of 25 µm and it was assumed that the value of $K_{H\text{-}P}$ was 0.24 MPa·m$^{1/2}$ [24]. It can be seen that strengthening of Ti by ECAP processing is quite adequately described by the Hall-Petch relationship. At the same time, with the grain sizes decreasing to 80 [4] and 40 nm (present work) the yield stress of Ti is somewhat lower than the predicted classic curve. Such a deviation from the Hall-Petch relationship and even its inverse dependence has been observed in other nanocrystalline metals, normally at grain sizes of 20–30 nm [25]. This is related, in the first place, to the realization of grain boundary sliding [25]. In developing these observations, a new aspect is the fact that in Ti the deviation from the classic Hall-Petch relationship is observed already at 80 nm, evidently, conditioned by the involvement of the grain sliding mechanism in the deformation, which can take place at room temperature [4]. However, this requires a more detailed study of the nature of this unusual phenomenon.

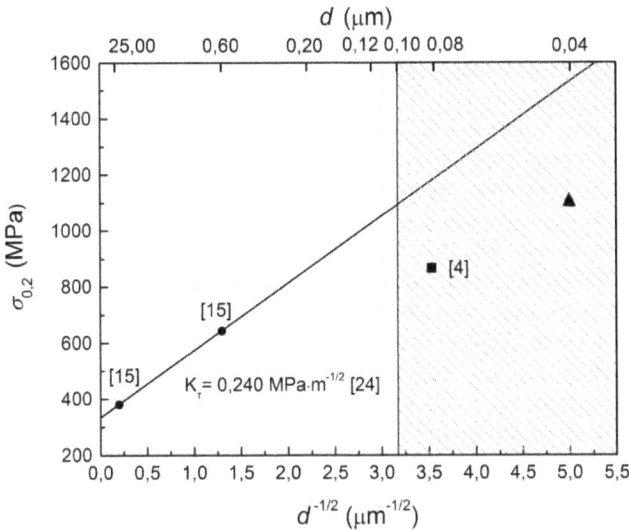

Figure 5. The relationship between grain size and yield stress ($\sigma_{0.2}$) and the experimental values for commercially pure (CP) Ti Grade 2 from earlier [4,15] and present works.

To provide insight into the influence of dispersed oxide particles on the strengthening mechanisms in CP Ti, the Orowan-Ashby equation was used to estimate their potential contribution to strength [26].

$$\Delta\sigma_{SD} = 0.16\frac{Gb}{\lambda}\ln\frac{r}{b} \tag{2}$$

where r is the average radius of nanoparticles (~10 nm), G is the shear modulus (41.4 GPa for Ti) and λ is the average distance between them,

$$\lambda = \left(\frac{4\pi r^3}{3f_o}\right)^{1/3} \tag{3}$$

where f is the volume fraction of oxide particles in the structure (about 0.1%). The Burgers vector was assumed to be 0.336 nm for $a + c$ slip systems along the piramidal plane, which were the most active ones in CP Ti during HPT [27]. Estimates show that the contribution of the oxides to the flow stress of nanocrystalline Ti is marginal (~4.7 MPa). According to the work [10], the changes in the concentration of oxygen near boundaries can also considerably affect the strength and ductility of UFG Ti after annealing at 623 K. From this work, it appears that the changes in the concentration of alloying elements near grain boundaries and the accompanying grain boundary segregations with their influence on the plastic deformation response should be also investigated in more detail during further study.

3. Experimental Section

CP Ti Grade 2 powder with the following chemical composition (Ti-base,-0.015H-0.052C-0.24O-0.3Fe-0.015N, wt.%) was used as the initial material for our studies. The powder was fabricated by cryogenic milling of commercial CP Ti in a liquid argon medium; the experimental details are available in the published literature [19]. The cryomilled powder was then subjected to preliminary compaction at a pressure of 1.95 GPa (described hereafter as pre-compaction), followed by SPD processing using high pressure torsion (HPT) at 573 K with 5 rotations under a pressure of 6 GPa. The samples were shaped as disks with a diameter of 10 mm and a thickness of 0.2 mm. Transmission electron microscopywas performed on a Philips CM 20 (TEM Philips is a trademark of Philips Electronic Instruments Corp., Mahwah, NJ, USA) operating at 200 kV with a condenser aperture with a 100 µm nominal diameter and a nominal beam diameter of 55 nm. TEM foils were prepared by twin-jet polishing in a solution of 5% perchloric acid and 95% methanol using a Struers Tenupol 5 electropolisher (Struers A/S, Ballerup, Denmark) at a temperature of −248 K. The operating voltage was 50 V. Observations were made in both bright and dark field imaging modes, and the selected area electron diffraction (SAED) patterns were recorded from the areas of interest using an aperture of 0.3 µm nominal diameter. The dislocation density was calculated on five bright and dark field TEM micrographs at magnification of 88,000 times with two tilting using intercept method, where the foil thickness was determined from intensity oscillations in the two-beam convergent beam electron diffraction (CBED) patterns. Microhardness was measured using a Micromet 5101 microhardness tester (BUEHLER LTD, Lake Bluff, IL, USA) at a load of 100–150g for 20 s.

4. Conclusions

For the first time, the possibility of producing in Grade 2 CP Ti a nanocrystalline structure with a grain size below 40 nm has been demonstrated. This is made possible due to high-strain-rate deformation at a cryogenic temperature in the powder milling and its subsequent consolidation by HPT at a high pressure (6 GPA) and a temperature of 573 K.

The discovered deviation of the experimental data on microhardness and yield stress from the typical Hall-Petch relationship for Ti Grade 2 extrapolated into the ultrafine-grained region may be associated with the involvement of the grain boundary sliding mechanism into deformation in nanostructured Ti already at a grain size below 100–80 nm.

Acknowledgments: The work has been done under the financial support of the Russian Federal Ministry for Education and Science and the U.S. National Science Foundation (grant number: DMR-1210437). R.Z. Valiev gratefully acknowledges the support from the Russian Federal Ministry for Education and Science (RZV Grant No 14.B25.31.0017), and I.P. Semenova acknowledges the support of the said Ministry within the scope of the basic part of the State Assignment. E.J. Lavernia would like to acknowledge the financial support from the US National Science Foundation under the grant number of NSF DMR-1210437.

Author Contributions: Irina Semenova: idea of the paper and major preparation of text; Ilana Timochina: transmission electron microscopy of samples; Rinat Islamgaliev: high pressure torsion (HPT) of samples and discussion of the experimental results; Enrique Lavernia: cryomilling of powders, the preliminary compaction of experimental samples and discussion of the experimental results; Ruslan Valiev: discussion of the experimental results.

Conflicts of Interest: The authors declare no conflict of interest.

References

1. Valiev, R.Z. Nanostructuring of metals by severe plastic deformation for advanced properties. *Nat. Mater.* **2004**, *3*, 511–516. [CrossRef] [PubMed]
2. Valiev, R.Z.; Islamgaliev, R.K.; Alexandrov, I.V. Bulk nanostructured materials from severe plastic deformation. *Prog. Mater. Sci.* **2000**, *45*, 103–189. [CrossRef]
3. Popov, A.A.; Pyshmintsev, I.Y.; Demakov, S.L.; Illarionov, A.G.; Lowe, T.C.; Valiev, R.Z. Structural and mechanical properties of nanocristalline Ti processed by severe deformation processing. *Scripta Mater.* **1997**, *37*, 1089–1094. [CrossRef]
4. Valiev, R.Z.; Sergueeva, A.V.; Mukherjee, A.K. The Effect of annealing on tensile deformation behavior of nanostructured SPD titanium. *Scripta Mater.* **2003**, *49*, 669–674. [CrossRef]
5. Purcek, G.; Yapici, G.G.; Karaman, I.; Maier, H.J. Effect of commercial purity levels on the mechanical properties of ultrafine-grained titanium. *Mater. Sci. Eng. A* **2011**, *528*, 2303–2308. [CrossRef]
6. Fan, Z.; Jiang, H.; Sun, X.; Song, J.; Zhang, X.; Xie, C. Microstructures and mechanical deformation behaviors of ultrafine-grained commercial pure (grade 3) Ti processed by two-step severe plastic deformation. *Mater. Sci. Eng. A* **2009**, *527*, 45–51. [CrossRef]

7. Zherebtsov, S.V.; Dyakonov, G.S.; Salem, A.A.; Malysheva, S.P.; Tselishchev, G.A.; Semiatin, S.L. Evolution of grain and subgrain structure during cold rolling of commercial-purity Ti. *Mater. Sci. Eng. A* **2011**, *528*, 3474–3479. [CrossRef]
8. Zherebtsov, S.V.; Dyakonov, G.S.; Salem, A.A.; Sokolenko, V.I.; Salishchev, G.A.; Semiatin, S.L. Formation of nanostructures in commercial-purity Ti via cryorolling. *Acta Mater.* **2013**, *61*, 1167–1178. [CrossRef]
9. Valiev, R.Z.; Alexandrov, I.V.; Enikeev, N.A.; Murashkin, M.Y.; Semenova, I.P. Towards enhancement of properties of UFG metals and alloys by grain boundary engineering using SPD processing. *Rev. Adv. Mater. Sci.* **2010**, *25*, 1–10.
10. Semenova, I.; Salimgareeva, G.; Costa, G.D.; Lefebvre, W.; Valiev, R. Enhanced strength and ductility of ultra-fine grained Ti processed by severe plastic deformation. *Adv. Eng. Mater.* **2010**, *12*, 803–807. [CrossRef]
11. Lapovok, R.; Tomusa, D.; Mang, J.; Estrin, Y.; Lowe, T.C. Evolution of nanoscale porosity during equal-channel angular pressing of titanium. *Acta Mater.* **2009**, *57*, 2909–2918. [CrossRef]
12. Zhao, X.; Fu, W.; Yang, X.; Langdon, T.G. Microstructure and properties of pure titanium processed by equal-channel angular pressing at room temperature. *Scripta Mater.* **2008**, *59*, 542–545. [CrossRef]
13. Stolyarov, V.V.; Zhu, Y.T.; Lowe, T.C.; Islamgaliev, R.K.; Valiev, R.Z. A two step SPD processing of ultrafine-grained Ti. *Nanostruct. Mater.* **1999**, *11*, 947–954. [CrossRef]
14. Valiev, R.Z.; Semenova, I.; Latysh, V.V.; Rack, H.; Lowe, T.C.; Petruzelka, J.; Dluhos, L.; Hrusak, D.; Sochova, J. Nanostructured Ti for biomedical applications. *Adv. Eng. Mater.* **2008**, *8*, B15–B17. [CrossRef]
15. Stolyarov, V.V.; Zhu, Y.T.; Alexandrov, I.V.; Lowe, T.C.; Valiev, R.Z. Influence of ECAP routes on the microstructure and properties of pure Ti. *Mater. Sci. Eng. A* **2001**, *299*, 59–67. [CrossRef]
16. Korznikov, A.V.; Safarov, I.M.; Laptionok, D.V.; Valiev, R.Z. Structure and properties of superfine-grained iron compacted out of ultradisperse powder. *Acta Metall. Mater.* **1991**, *39*, 3193–3197. [CrossRef]
17. Valiev, R.Z. Processing of nanocrystalline materials by severe plastic deformation consolidation. In *Synthesis and Processing of Nanocrystalline Powder*; Bourell, D.L., Ed.; The Minerals, Metals and Materials Society: Warrendale, PA, USA, 1996; pp. 153–161.
18. Wen, H.; Zhao, Y.; Liu, Y.; Torera, O.; Nesterov, K.M.; Islamgaliev, R.K.; Valiev, R.Z.; Lavernia, E.J. High-pressure torsion-induced grain growth and detwinning in cryomilled Cu powders. *Phil. Mag.* **2010**, *90*, 4541–4550. [CrossRef]
19. Witkin, D.B.; Lavernia, E.J. Synthesis and mechanical behavior of nanostructured materials via cryomilling. *Prog. Mater. Sci.* **2006**, *51*, 1–60. [CrossRef]
20. Chen, D.-J.; Mayo, M.J. Densification and grain growth of ultra fine 3 mol % Y_2O_3-ZrO_2 ceramics. *Nanostruct. Mater.* **1993**, *2*, 469–478. [CrossRef]
21. Rabe, T.; Wasche, R. Sintering behaviour of nanocrystalline titanium nitride powders. *Nanostruct. Mater.* **1995**, *6*, 357–360. [CrossRef]
22. Meyers, M.A.E.; Chawla, K.K. *Mechanical Behavior of Materials*, 2nd ed.; University Press: Cambridge, UK, 2009.
23. Ahn, S.H.; Chun, Y.B.; Yu, S.H.; Kim, K.H.; Hwang, S.K. Microstructural refinement and deformation mode of Ti under cryogenic channel die compression. *Mater. Sci. Eng. A* **2010**, *528*, 165–171. [CrossRef]
24. Stanford, N.; Carlson, U.; Barnett, M.R. Deformation twinning and the Hall-Petch relation in commercial purity Ti. *Metall Mater. Trans. A* **2008**, *39*, 934–944. [CrossRef]
25. Conrad, H.; Narayan, J. On the grain size softening in nanocrystalline materials. *Scripta Mater.* **2000**, *42*, 1025–1030. [CrossRef]
26. Frost, H.J.; Ashby, M.F. *The Plasticity and Creep of Metals and Ceramics*; Pergamon Press: Oxford, UK, 1982.
27. Sitdikov, V.D.; Alexandrov, I.V.; Bonarski, J.T. X-ray analysis of α and ω-phases of Ti, subjected to high-pressure torsion. *Mat. Sci. Eng.* **2011**, *667–669*, 187–192.

MDPI

Article

Manufacturing Ultrafine-Grained Ti-6Al-4V Bulk Rod Using Multi-Pass Caliber-Rolling

Taekyung Lee [1], Donald S. Shih [2], Yongmoon Lee [3] and Chong Soo Lee [3,*]

[1] Department of Mechanical Engineering, Northwestern University, Evanston, IL 60208, USA; taekyung.lee@northwestern.edu

[2] Boeing Research & Technology, St. Louis, MO 63166, USA; donald.s.shih@boeing.com

[3] Graduate Institute of Ferrous Technology, Pohang University of Science and Technology (POSTECH), Pohang 790-784, Korea; ymlee0725@postech.ac.kr

* Author to whom correspondence should be addressed; cslee@postech.ac.kr; Tel.: +82-54-279-9001; Fax: +82-54-279-9099.

Academic Editor: Heinz Werner Höppel

Received: 11 March 2015; Accepted: 30 April 2015; Published: 15 May 2015

Abstract: Ultrafine-grained (UFG) Ti-6Al-4V alloy has attracted attention from the various industries due to its good mechanical properties. Although severe plastic deformation (SPD) processes can produce such a material, its dimension is generally limited to laboratory scale. The present work utilized the multi-pass caliber-rolling process to fabricate Ti-6Al-4V bulk rod with the equiaxed UFG microstructure. The manufactured alloy mainly consisted of alpha phase and showed the fiber texture with the basal planes parallel to the rolling direction. This rod was large enough to be used in the industry and exhibited comparable tensile properties at room temperature in comparison to SPD-processed Ti-6Al-4V alloys. The material also showed good formability at elevated temperature due to the occurrence of superplasticity. Internal-variable analysis was carried out to measure the contribution of deformation mechanisms at elevated temperatures in the manufactured alloy. This revealed the increasing contribution of phase/grain-boundary sliding at 1073 K, which explained the observed superplasticity.

Keywords: multi-pass caliber-rolling; grain refinement; internal-variable theory; Ti-6Al-4V

1. Introduction

Titanium and its alloys have attracted attention from various fields such as structural-material, biomedical, munitions, and information-technology industries. In particular, Ti-6Al-4V alloy has been a key material in aerospace industries since it was developed in 1954 [1]. The alloy possesses a superior strength-to-weight ratio that increases the fuel efficiency of rocket and aircraft. It also exhibits excellent corrosion resistance and mechanical stability at various temperatures. Finally, the alloy has good formability at elevated temperatures after applying a certain thermomechanical process due to the superplasticity [2–4].

Many researchers have focused on the fact that grain refinement can improve mechanical properties of titanium alloys. The increasing fraction of grain boundaries act as barriers against dislocation slip, leading to grain-boundary strengthening. In addition, the grain refinement provides the increasing sources of grain-boundary sliding and hence induces the superplastic behavior at elevated temperatures. It is thus natural that ultrafine-grained (UFG) titanium alloys have been actively studied for decades [5–9]. To attain the UFG structure, most of previous works have utilized a severe plastic deformation (SPD) process, such as equal-channel angular pressing (ECAP) and high-pressure torsion (HPT) [10]. However, the manufactured samples were generally 10s of centimeters in length, which were limited to laboratory scale.

To produce a UFG bulk rod applicable to the industries, the authors have introduced a multi-pass caliber-rolling process as an alternative to the conventional SPD processes. Caliber-rolling machine includes several calibers with various sizes and shapes (e.g., oval and circular) in its rolls by which a multi-axial deformation is imposed on a workpiece during the process. Since Kimura *et al.* [11] reported the considerable mechanical improvement in a caliber-rolled low-alloy steel, related studies have been actively carried out with various materials [12–19]. Nevertheless, studies on caliber-rolled titanium alloys have just begun in spite of its importance both in academia and industry [20–22]. A UFG bulk rod was successfully manufactured by caliber-rolling Ti-6Al-4V alloy in this work. The researchers investigated microstructures and tensile properties of the manufactured material and discussed mechanisms of grain refinement and superplasticity.

2. Materials and Methods

Ti-6Al-4V rod was machined with a diameter of 28 mm and a length of 150 mm. The head part was made to be conical to insert a material into a pair of calibers. The beta-transus temperature of this alloy was reported to be 1268 K [9]. The Ti-6Al-4V rod was solution-treated at 1323 K for 2 h followed by quenching in a water bath to induce a fine lath structure. This alloy was then soaked in a furnace at 1073 K for 1 h and caliber-rolled in the ambient atmosphere. The caliber-rolling process used in this work consists of six deformation passes. First, third, and fifth calibers are oval-shaped, while second, fourth, and sixth calibers are circular. The sample was inserted in a caliber, rotated by 90 degrees, and then immediately inserted in the next caliber for each deformation path, as illustrated in Figure 1. There was a single reheating process at 1073 K for 2 min after the fourth deformation pass. The sample was air-cooled after the sixth deformation pass. The total reduction of area was determined to be 85%.

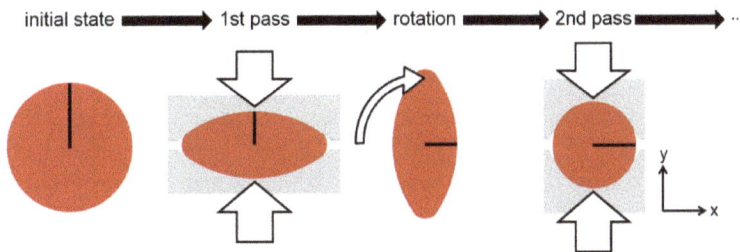

Figure 1. Schematic illustration of multi-pass caliber-rolling process.

Disc samples with a diameter of 3 mm were obtained from the caliber-rolled alloy, and then mechanically thinned with 240-grit SiC papers to a thickness less than 200 μm. Afterwards, they were jet-polished at 22 V and 265 K in a solution containing 40 mL of $HClO_4$, 240 mL of 2-butoxy ethanol, and 400 mL of methanol. Transmission electron microscope (TEM) was used to observe microstructures at 200 kV with JEM-2100F FE-TEM machine (JEOL, Tokyo, Japan). Grain size was measured from the image using the linear intercept method [23]. Meanwhile, other discs were mirror-polished with 1 μm alumina powder and 0.25 μm colloidal silica for electron backscatter diffraction (EBSD) analysis. The analysis was performed at 20 kV with Quanta 3D FEG machine (FEI, Hillsboro, OR, USA).

Tensile properties were measured using rod-type specimens whose gauge length and diameter were 6 and 3 mm, respectively. Instron 8862 machine (INSTRON, Norwood, MA, USA) was used for both room- and high-temperature tensile tests. Room-temperature tensile test was carried out at a strain rate of 5×10^{-3} s^{-1} with an extensometer to obtain reliable data. High-temperature tensile test was conducted in a halogen furnace at two temperature conditions (873 K and 1073 K) and two strain rates (5×10^{-3} s^{-1} and 5×10^{-4} s^{-1}). Each sample was heated for 10 min before commencing the test. Load-relaxation test (LRT) was conducted in a similar manner to the high-temperature tensile test, except that the samples were deformed up to a true strain of 0.2 and then hold to investigate the load relaxation behavior. The obtained LRT data were converted into stress-strain rate relationship based on the following equation [24]:

$$\sigma = P(L_0 + X - P/K)/A_0 L_0 \tag{1a}$$

$$\dot{\varepsilon} = -(dP/dt)(L_O + X - P/K) \tag{1b}$$

$$K^{-1} \approx C_m + L_0/A_0 E \tag{1c}$$

where P is load, L_o is the gauge length, A_o is the cross-sectional area of gauge region, X is the displacement, C_m is the elastic compliance of testing machine, and E is Young's modulus of Ti-6Al-4V alloy. Strain-rate-jump test (SRJT) was performed at 1073 K, during which a strain rate changed from 5×10^{-4} s^{-1} to 5×10^{-3} s^{-1} at a true strain of 0.6. Strain-rate sensitivity (m) was determined as follows:

$$m = \partial \log \infty / \partial \log \dot{\varepsilon} \tag{2}$$

3. Results

Figure 2 demonstrates the manufactured caliber-rolled Ti-6Al-4V bulk rod. The length and diameter of the rod was approximately 1200 and 10 mm, respectively. It is also noted that the length can be even increased by tailoring the dimension of initial material. Such a large dimension enables the caliber-rolled rod to be directly used in the industry. Indeed, the authors fabricated a dental implant fixture with this material, which exhibited satisfying mechanical strength and fatigue resistance both in ambient atmosphere and simulated body fluid [21].

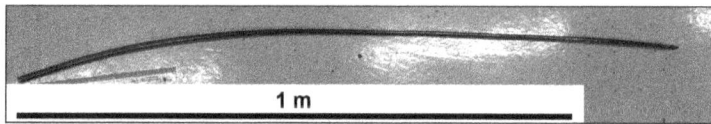

Figure 2. Ti-6Al-4V bulk rod manufactured by the multi-pass caliber-rolling process.

The grain structure of caliber-rolled rod was observed by TEM analysis as shown in Figure 3. It is obvious that the rod possessed the UFG structure with a grain size of 0.2 ± 0.05 μm. Such a strong grain refinement has thus far been accomplished by SPD processes. For example, Ko *et al.* [25] achieved a similar UFG structure with a mean grain size of ~0.3 μm after applying four-pass ECAP deformation at 873 K; however, the length of fabricated sample was much shorter (80 mm) compared to the present UFG rod.

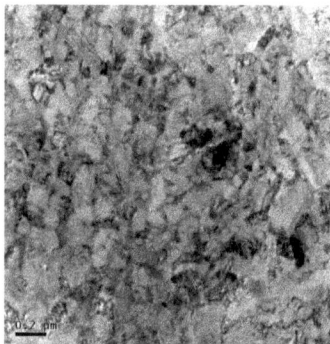

Figure 3. Transmission electron microscope (TEM) micrograph of caliber-rolled Ti-6Al-4V rod. The image was taken perpendicular to the rolling direction (RD).

Figure 4 presents the EBSD results for the investigated materials. The solution-treated microstructure consisted of martensitic laths as intended. Beta phase was not confirmed in this alloy, as reported in the literature [26,27]. Two types of martensitic laths were observed; coarse primary laths were formed first, and then fine secondary laths were generated between the primary laths [28]. The latter occupies the most area in the solution-treated alloy, whose lath thickness is less than 5 μm. The similar lath structure with no presence of beta phase was confirmed after the heating step prior to the caliber-rolling (*i.e.*, 1073 K for 1 h), although the thickness was increased to ~10 μm. The beta fraction increased to 6% after the caliber-rolling. Figure 5 shows two types of beta constituents; most of them exist as a form of nano-sized beta precipitations, while fragmented beta lamellae are also observed. The beta lamellae were formed parallel to the RD and the thickness was measured to be ~0.2 μm or less. Chao *et al.* [27] attributed this type of phase transformation to adiabatic heating and strain-induced transformation. A fraction of high-angle grain boundaries of 0.6–0.8 was confirmed in Ti-6Al-4V alloys groove-rolled at 923–1023 K [22]. Similar results are expected in the present material as both rolling processes have the similar deformation mechanism.

The confidence index (CI) of caliber-rolled rod was too low to provide meaningful microstructural information. Tirumalasetty *et al.* [29] ascribed the low CI in UFG alloys to distorted Kikuchi patterns in region with high dislocation density. Alternatively, the EBSD analysis was conducted after annealing the caliber-rolled alloy at 873 K for 1 h, followed by water-quenching; such a condition was reported to minimize recrystallization and maintain the texture in this material [30]. Figure 4c,d demonstrate two types of grains: fine and coarse-grain groups. The fine grains are closer to the original microstructure of caliber-rolled alloy because the clear image and high CI of coarse grains imply the formation of microstructure during the subsequent annealing process. It should be noted that the fine grains are equiaxed in both planes, supporting the conclusion from TEM observation that the caliber-rolling gave rise to the equiaxed UFG structure for the present material.

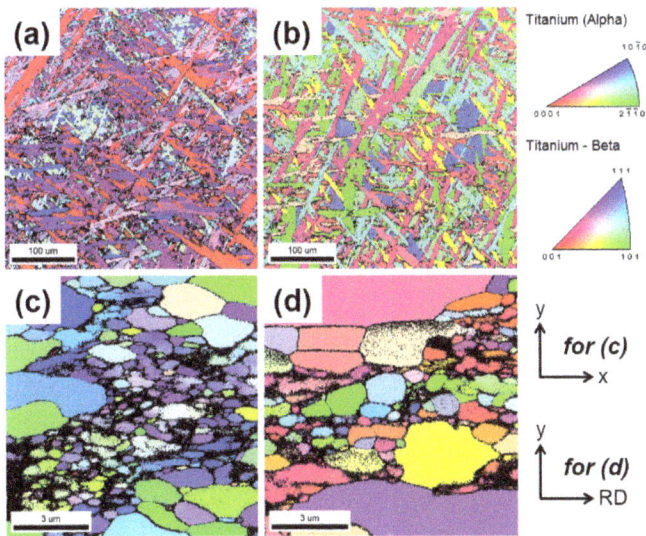

Figure 4. Electron backscatter diffraction (EBSD) pole figure map of the investigated Ti-6Al-4V rod: (a) solution-treated; (b) heat-treated prior to the caliber-rolling; and (c,d) caliber-rolled and annealed alloys. Figure 4c is a plane perpendicular to the RD, while Figure 4d is parallel to the RD. The black dots indicate the area of confidence index (CI) ≤ 0.1. The *x* and *y* axes are demonstrated in Figure 1.

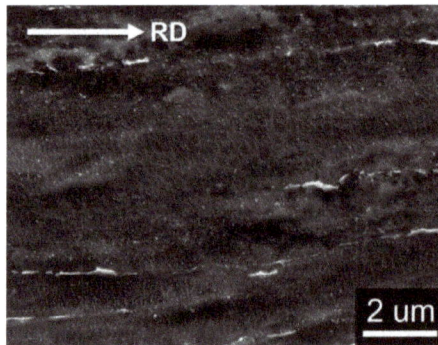

Figure 5. SEM micrograph of the caliber-rolled Ti-6Al-4V rod. The dark and bright areas indicate alpha and beta phases, respectively. The image was taken parallel to the RD marked as the arrow in the micrograph.

Figure 6 presents the texture of the caliber-rolled rod in the form of RD inverse pole figure. The alloy exhibited the fiber texture with the basal planes parallel to the RD. The fraction of beta phase is small enough to be neglected. Narayana Murty *et al.* [22] recently reported the similar texture in Ti-6Al-4V alloys groove-rolled at 873–1023 K, although their texture showed stronger orientation along <10-10> and <2-1-10> directions. According to the work [22], the relatively randomized texture observed in the present alloy may be attributed to higher deformation temperature related to martensite decomposition and phase transformation during the deformation process as well as the annealing treatment.

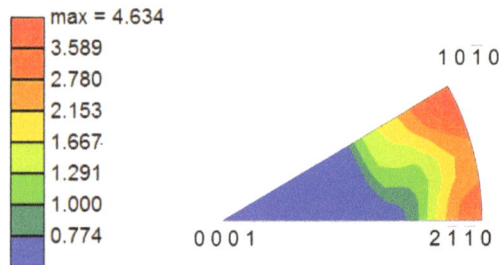

Figure 6. RD inverse pole figure for the caliber-rolled Ti-6Al-4V rod obtained by EBSD.

The manufactured UFG rod can be utilized in two ways in the industry. First, the material can be directly machined to be used for biomedical products, such as dental implant fixture, bone screw, bone plate, and micro-drill [1]. Room-temperature tensile properties are important in this case to ensure the resistance to fatigue fracture for biomedical uses. Second, the rod can be further processed at elevated temperature for automobile and aerospace industries, which requires the evaluation of high-temperature mechanical properties. Therefore, the tensile properties of the caliber-rolled Ti-6Al-4V alloy were investigated at both room and elevated temperatures.

Figure 7 shows room-temperature tensile properties of the caliber-rolled rod as well as UFG Ti-6Al-4V alloys in the literature [5–8]. The caliber-rolled rod provided the high yield stress (YS = 1345 MPa) and ultimate tensile stress (UTS = 1425 MPa) due to its UFG structure and resultant grain-boundary strengthening. The tensile properties of the manufactured alloy was compared with SPD-processed UFG Ti-6Al-4V alloys in terms of the product of strength and elongation; such an approach has been widely used in the field of structural materials as strength increases in sacrifice of elongation in many cases [31]. The UFG bulk Ti-6Al-4V rod fabricated in this work exhibited the value of 18,525 MPa%, as shown in Figure 7b, which was comparable with most of UFG Ti-6Al-4V alloys in the literature.

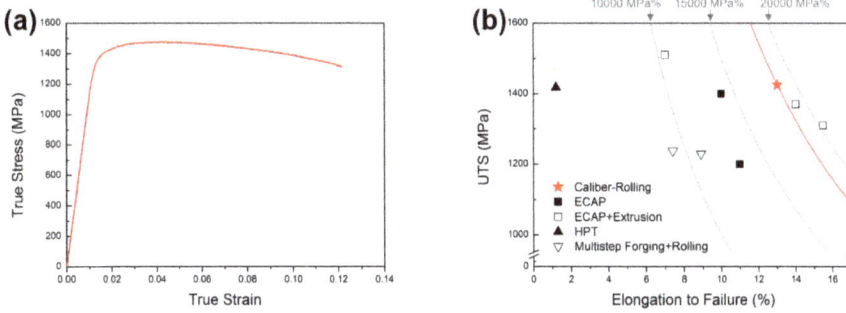

Figure 7. (**a**) True stress-strain curve of caliber-rolled Ti-6Al-4V rod at room temperature and (**b**) comparison of the tensile properties of various ultrafine-grained (UFG) Ti-6Al-4V alloys [5–8].

Figure 8a shows flow curves of the caliber-rolled alloy at elevated temperatures. Increasing deformation temperature or decreasing strain rate increased the ductility in sacrifice of strength. High-temperature deformation mechanisms at each condition were discussed in Section 4 on the basis of the internal-variable theory. It is of particular note that the caliber-rolled rod deformed at 1073 K and 5×10^{-4} s^{-1} exhibited the different characteristics than the others. The sample recorded a total elongation of 967% as shown in the inset of Figure 8a, whose flow curve showed a long plateau as generally found in superplastic materials. In addition, the SRJT results shown in Figure 8b provided a strain-rate sensitivity of 0.44 for the caliber-rolled Ti-6Al-4V at 1073 K. A material with a strain-rate sensitivity of 0.3–0.8 is considered to possess superplasticity [32]. All of these factors suggest the superplastic behavior of the present bulk rod at the high temperature, which will be useful in the related industries.

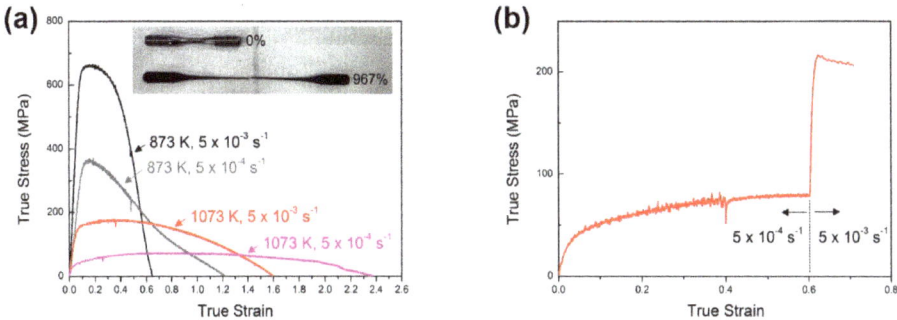

Figure 8. High-temperature tensile properties of caliber-rolled Ti-6Al-4V rod: (**a**) true stress-strain curve at elevated temperatures and (**b**) SRJT results obtained at 1073 K. The inset in Figure 8a compares the undeformed specimen and the sample deformed at 1073 K and 5×10^{-4} s^{-1}.

4. Discussion

The caliber-rolling process gave rise to the strong grain refinement by which the grain size became similar to those refined by the SPD processes. Such a microstructural evolution can be understood in terms of the dynamic globularization [33–36]. A groove formed at the phase/grain boundaries splits a platelet into several pieces by deepening along the boundary in the platelet. The broken-up lamellae are transformed into globular grains to stabilize the surface energy.

Two factors contributed to the effective dynamic globularization and resultant grain refinement in the caliber-rolled Ti-6Al-4V alloy. First, the present alloy consisted of fine martensitic laths prior to the caliber-rolling process. According to the literature [27,37,38], such a microstructure is beneficial for the grain refinement through dynamic globularization because an initial lamellar thickness directly affects a final grain size. Second, the globularizing fraction increases with increase in an applied strain following an Avrami-type equation in Ti-6Al-4V [39]. The authors have proven that the caliber-rolling imposes more than twice as high strain than a conventional rolling with the same reduction of area [30]. In this work, the equivalent strain was determined to be 0.7, 1.2, 1.9, 2.4, 3.3, and 4.0 after the one- to six-pass caliber-rolling in a two-phase titanium alloy. This is attributed to the redundant strain accumulating without the volume change of workpiece [40].

Superplastic behavior was confirmed in the manufactured alloy at 1073 K and 5×10^{-4} s^{-1}. The high-temperature deformation data were interpreted on the basis of internal-variable theory to investigate the deformation mechanisms. Ha and Chang [41] suggested this theory to measure the contribution of grain matrix deformation (GMD) and phase/grain-boundary sliding (P/GBS) to deformation behavior at elevated temperatures. In the internal-variable theory, stress is composed of internal stress of long-range dislocation interactions (σ^I) and friction stress of short-range dislocation-lattice interactions (σ^F). Strain rate consists of the rate of internal strain ($\dot{\iota}$), non-recoverable plastic strain ($\dot{\alpha}$), and P/GBS ($\dot{\eta}$) strain. Among these factors, the friction stress and internal strain rate are negligible under the present conditions, providing stress and strain rate as follows:

$$\sigma = \sigma^I + \sigma^F \approx \sigma^I \tag{3a}$$

$$\dot{\varepsilon} = \dot{\iota} + \dot{\alpha} + \dot{\eta} \approx \dot{\alpha} + \dot{\eta} \tag{3b}$$

The internal stress, non-recoverable plastic strain rate, and P/GBS strain rate at a deformation temperature of T are determined from the following relations:

$$(\sigma^*/\sigma^I) = \exp(\dot{\alpha}^*/\dot{\alpha})^p \tag{4a}$$

$$\dot{\alpha}^* = v^I (\sigma^*/G)^{n(I)} \exp(-Q^I/RT) \tag{4b}$$

$$(\dot{\eta}/\dot{\eta}_o) = [(\sigma - \Sigma_\eta)/\Sigma_\eta]^{1/M} \tag{4c}$$

$$\dot{\eta}_o = v^\eta (\Sigma_\eta/\mu^\eta)^{n(\eta)} \exp(-Q^\eta/RT) \tag{4d}$$

Here, σ^* and Σ_η are stress for GMD and P/GBS, respectively. $\dot{\alpha}^*$ and $\dot{\eta}_o$ are their conjugate reference strain rates. v^I and v^η are the jump frequency for dislocations. Q^I and Q^η are the activation energy for GMD and P/GBS, respectively. R is the gas constant and other parameters are material constants [4].

Figure 9 shows LRT results and corresponding internal-variable analysis. The GMD curve deviated from the experimental data at lower strain rates of Figure 9a and almost entire range of Figure 9b. These deviations were corrected by the P/GBS curve. It is thus concluded that P/GBS was not activated at 873 K and a strain rate of 5×10^{-3} s^{-1}, whereas both mechanisms contributed to the high-temperature deformation in the other cases. The relative contribution of each mechanism was quantified on the basis of the LRT data. At 873 K and a strain rate of 5×10^{-4} s^{-1}, GMD mainly contributed to the deformation behavior (92% for GMD and 8% for P/GBS). Similar results were obtained at 1073 K and a strain rate of 5×10^{-3} s^{-1} (85% for GMD and 15% for P/GBS) from the extrapolated data in Figure 9b. These results explain why the superplasticity was not observed under the three conditions mentioned above.

On the other hand, the contribution of P/GBS significantly increased at 1073 K and a strain rate of 5×10^{-4} s^{-1} (55% for GMD and 45% for P/GBS). This is rationalized in light of the decreasing activation energy for deformation with increasing temperature and decreasing strain rate. The authors have reported that the activation energy decreased to that for interphase/grain-boundary diffusion under such conditions, resulting in the increasing contribution of P/GBS [4]. This conclusion is also supported by the fact that the friction stress for P/GBS decreased as the deformation temperature increased; the value of log Σ_η was −2.7 at 873 K and −5.7 at 1073 K. This is in good accord with the superplastic behavior observed at 1073 K and 5×10^{-4} s^{-1}.

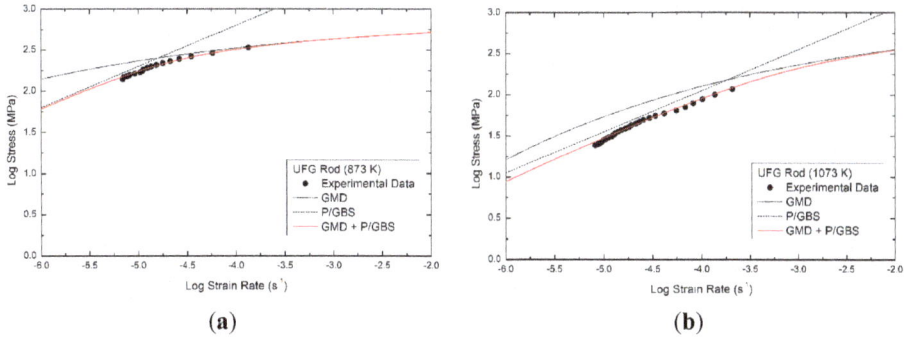

Figure 9. Internal-variable analysis of caliber-rolled Ti-6Al-4V rod at (**a**) 873 K and (**b**) 1073 K.

5. Conclusions

In this work, the multi-pass caliber-rolling process successfully manufactured Ti-6Al-4V bulk rod with the UFG microstructure. The dimension of the manufactured rod (1200 mm in length) was significantly larger than most UFG Ti-6Al-4V samples fabricated by SPD processes and large enough to be directly used in the industry. The length can be even increased by tailoring the dimension of initial material. The alloy consisted of equiaxed ultrafine grains with a mean size of 0.2 μm. Such an effective grain refinement originated from the fine lath structure in the initial material and redundant strain accumulating without the volume change of workpiece. A small amount of beta particles was formed during the caliber-rolling process due to adiabatic heating and strain-induced phase transformation. The caliber-rolled rod showed the fiber texture with the basal planes parallel to the RD. The grain refinement through the caliber-rolling affected the grain-boundary strengthening at room temperature and superplastic behavior at high temperature. The product of strength and elongation of the alloy at room temperature was calculated to be 18525 MPa%, which was similar to the values of most SPD-processed UFG Ti-6Al-4V alloys in the literature. The manufactured rod exhibited the superplastic behavior at 1073 K and 5×10^{-4} s^{-1}. The internal-variable analysis revealed the increasing P/GBS contribution to deformation under these conditions, while GMD controlled the deformation at the lower temperature and/or higher strain rate.

Acknowledgments: The authors gratefully acknowledge the financial support from The Boeing Company for the present research.

Author Contributions: T. Lee and C.S. Lee conceived and designed the experiments; T. Lee performed all experiments except for EBSD, analyzed the data, and wrote the paper; D.S. Shih and C.S. Lee guided the direction of the work and contributed reagents/materials/analysis tools; Y. Lee carried out the entire EBSD analyses; all authors participated in discussions.

Conflicts of Interest: The authors declare no conflict of interest.

References

1. Lee, Y.T. *Titanium*; Korea Metal Journal: Seoul, Korea, 2009.

2. Park, C.H.; Ko, Y.G.; Park, J.-W.; Lee, C.S. Enhanced superplasticity utilizing dynamic globularization of Ti-6Al-4V alloy. *Mater. Sci. Eng. A* **2008**, *496*, 150–158. [CrossRef]

3. Semiatin, S.L.; Fagin, P.N.; Betten, J.F.; Zane, A.P.; Ghosh, A.K.; Sargent, G.A. Plastic flow and microstructure evolution during low-temperature superplasticity of ultrafine Ti-6Al-4V sheet material. *Metall. Mater. Trans. A* **2010**, *41*, 499–512. [CrossRef]

4. Lee, T.; Kim, J.H.; Semiatin, S.L.; Lee, C.S. Internal-variable analysis of high-temperature deformation behavior of Ti-6Al-4V: A comparative study of the strain-rate-jump and load-relaxation tests. *Mater. Sci. Eng. A* **2013**, *562*, 180–189. [CrossRef]

5. Mishra, R.S.; Stolyarov, V.V.; Echer, C.; Valiev, R.Z.; Mukherjee, A.K. Mechanical behavior and superplasticity of a severe plastic deformation processed nanocrystalline Ti-6Al-4V alloy. *Mater. Sci. Eng. A* **2001**, *298*, 44–50. [CrossRef]

6. Salishchev, G.A.; Galeyev, R.M.; Valiakhmetov, O.R.; Safiullin, R.V.; Lutfullin, R.Y.; Senkov, O.N.; Froes, F.H.; Kaibyshev, O.A. Development of Ti-6Al-4V sheet with low temperature superplastic properties. *J. Mater. Process. Technol.* **2001**, *116*, 265–268. [CrossRef]

7. Semenova, I.P.; Raab, G.I.; Saitova, L.R.; Valiev, R.Z. The effect of equal-channel angular pressing on the structure and mechanical behavior of Ti-6Al-4V alloy. *Mater. Sci. Eng. A* **2004**, *387–389*, 805–808. [CrossRef]

8. Saitova, L.R.; Höppel, H.W.; Göken, M.; Semenova, I.P.; Valiev, R.Z. Cyclic deformation behavior and fatigue lives of ultrafine-grained Ti-6AL-4V ELI alloy for medical use. *Int. J. Fatigue* **2009**, *31*, 322–331. [CrossRef]

9. Park, C.H.; Kim, J.H.; Yeom, J.-T.; Oh, C.-S.; Semiatin, S.L.; Lee, C.S. Formation of a submicrocrystalline structure in a two-phase titanium alloy without severe plastic deformation. *Scr. Mater.* **2013**, *68*, 996–999. [CrossRef]

10. Azushima, A.; Kopp, R.; Korhonen, A.; Yang, D.Y.; Micari, F.; Lahoti, G.D.; Groche, P.; Yanagimoto, J.; Tsuji, N.; Rosochowski, A.; *et al.* Severe plastic deformation (SPD) processes for metals. *CIRP Ann. Manuf. Technol.* **2008**, *57*, 716–735. [CrossRef]

11. Kimura, Y.; Inoue, T.; Yin, F.; Tsuzaki, K. Inverse temperature dependence of toughness in an ultrafine grain-structure steel. *Science* **2008**, *320*, 1057–1060. [CrossRef] [PubMed]

12. Yin, F.; Hanamura, T.; Inoue, T.; Nagai, K. Fiber texture and substructural features in the caliber-rolled low-carbon steels. *Metall. Mater. Trans. A* **2004**, *35*, 665–677. [CrossRef]

13. Torizuka, S.; Ohmori, A.; Narayana Murty, S.V.S.; Nagai, K. Effect of strain on the microstructure and mechanical properties of multi-pass warm caliber rolled low carbon steel. *Scr. Mater.* **2006**, *54*, 563–568. [CrossRef]

14. Inoue, T.; Yin, F.; Kimura, Y. Strain distribution and microstructural evolution in multi-pass warm caliber rolling. *Mater. Sci. Eng. A* **2007**, *466*, 114–122. [CrossRef]

15. Lee, T.; Park, C.H.; Lee, D.-L.; Lee, C.S. Enhancing tensile properties of ultrafine-grained medium-carbon steel utilizing fine carbides. *Mater. Sci. Eng. A* **2011**, *528*, 6558–6564. [CrossRef]

16. Lee, T.; Koyama, M.; Tsuzaki, K.; Lee, Y.-H.; Lee, C.S. Tensile deformation behavior of Fe–Mn–C TWIP steel with ultrafine elongated grain structure. *Mater. Lett.* **2012**, *75*, 169–171. [CrossRef]

17. Chun, Y.S.; Lee, J.; Bae, C.M.; Park, K.-T.; Lee, C.S. Caliber-rolled TWIP steel for high-strength wire rods with enhanced hydrogen-delayed fracture resistance. *Scr. Mater.* **2012**, *67*, 681–684. [CrossRef]

18. Lee, T.; Park, C.H.; Lee, S.-Y.; Son, I.-H.; Lee, D.-L.; Lee, C.S. Mechanisms of tensile improvement in caliber-rolled high-carbon steel. *Met. Mater. Int.* **2012**, *18*, 391–396. [CrossRef]

19. Doiphode, R.; Kulkarni, R.; Narayana Murty, S.V.S.; Prabhu, N.; Prasad Kashyap, B. Effect of severe caliber rolling on superplastic properties of Mg-3Al-1Zn (AZ31) alloy. *Mater. Sci. Forum* **2012**, *735*, 327–331. [CrossRef]

20. Lee, T.; Heo, Y.-U.; Lee, C.S. Microstructure tailoring to enhance strength and ductility in Ti–13Nb–13Zr for biomedical applications. *Scr. Mater.* **2013**, *69*, 785–788. [CrossRef]

21. Jung, H.S.; Lee, T.; Kwon, I.K.; Kim, H.S.; Hahn, S.K.; Lee, C.S. Surface modification of multi-pass caliber-rolled Ti alloy with dexamethasone-loaded graphene for dental applications. *ACS Appl. Mater. Interfaces* **2015**, *7*, 9598–9607. [CrossRef]

22. Narayana Murty, S.V.S.; Nayan, N.; Kumar, P.; Narayanan, P.R.; Sharma, S.C.; George, K.M. Microstructure-texture-mechanical properties relationship in multi-pass warm rolled Ti-6Al-4V alloy. *Mater. Sci. Eng. A* **2014**, *589*, 174–181. [CrossRef]

23. Dieter, G.E. *Mechanical Metallurgy*; Metric, S.I., Ed.; McGraw-Hill: London, UK, 1988.

24. Lee, D.; Hart, E. Stress relaxation and mechanical behavior of metals. *Metall. Mater. Trans. B* **1971**, *2*, 1245–1248. [CrossRef]

25. Ko, Y.G.; Lee, C.S.; Shin, D.H.; Semiatin, S.L. Low-temperature superplasticity of ultra-fine-grained Ti-6Al-4V processed by equal-channel angular pressing. *Metall. Mater. Trans. A* **2006**, *37*, 381–391. [CrossRef]

26. Matsumoto, H.; Bin, L.; Lee, S.-H.; Li, Y.; Ono, Y.; Chiba, A. Frequent occurrence of discontinuous dynamic recrystallization in Ti-6Al-4V Alloy with α' martensite starting microstructure. *Metall. Mater. Trans. A* **2013**, *44*, 3245–3260. [CrossRef]

27. Chao, Q.; Hodgson, P.D.; Beladi, H. Ultrafine grain formation in a Ti-6Al-4V alloy by thermomechanical processing of a martensitic microstructure. *Metall. Mater. Trans. A* **2014**, *45*, 2659–2671. [CrossRef]

28. Williams, J.C.; Taggart, R.; Polonis, D.H. The morphology and substructure of Ti-Cu martensite. *Metall. Mater. Trans. B* **1970**, *1*, 2265–2270. [CrossRef]

29. Tirumalasetty, G.K.; van Huis, M.A.; Kwakernaak, C.; Sietsma, J.; Sloof, W.G.; Zandbergen, H.W. Unravelling the structural and chemical features influencing deformation-induced martensitic transformations in steels. *Scr. Mater.* **2014**, *71*, 29–32. [CrossRef]

30. Lee, C.S.; Lee, T.; Kim, J.H.; Park, C.H. brication of ultrafine-grained Ti-6Al-4V bulk sheet/rod for related industries and their mechanical characteristics. In Proceedings of the 8th Pacific Rim International Conference on Advanced Materials and Processing (PRICM 8), Waikoloa, HI, USA, 4–9 August 2013.

31. Bouaziz, O.; Allain, S.; Scott, C.P.; Cugy, P.; Barbier, D. High manganese austenitic twinning induced plasticity steels: A review of the microstructure properties relationships. *Curr. Opin. Solid State Mater. Sci.* **2011**, *15*, 141–168. [CrossRef]

32. Reed-Hill, R.E.; Cribb, W.R.; Monteiro, S.N. Concerning the analysis of tensile stress-strain data using log $d\sigma/d\varepsilon$ *versus* log σ diagrams. *Metall. Mater. Trans. B* **1973**, *4*, 2665–2667. [CrossRef]

33. Seshacharyulu, T.; Medeiros, S.C.; Morgan, J.T.; Malas, J.C.; Frazier, W.G.; Prasad, Y.V.R.K. Hot deformation and microstructural damage mechanisms in extra-low interstitial (ELI) grade Ti-6Al-4V. *Mater. Sci. Eng. A* **2000**, *279*, 289–299. [CrossRef]

34. Stefansson, N.; Semiatin, S.L. Mechanisms of globularization of Ti-6Al-4V during static heat treatment. *Metall. Mater. Trans. A* **2003**, *34*, 691–698. [CrossRef]

35. Zherebtsov, S.; Murzinova, M.; Salishchev, G.; Semiatin, S.L. Spheroidization of the lamellar microstructure in Ti-6Al-4V alloy during warm deformation and annealing. *Acta Mater.* **2011**, *59*, 4138–4150. [CrossRef]

36. Park, C.H.; Won, J.W.; Park, J.-W.; Semiatin, S.L.; Lee, C.S. Mechanisms and kinetics of static spheroidization of hot-worked Ti-6Al-2Sn-4Zr-2Mo-0.1Si with a lamellar microstructure. *Metall. Mater. Trans. A* **2012**, *43*, 977–985. [CrossRef]

37. Park, C.H.; Park, J.-W.; Yeom, J.-T.; Chun, Y.S.; Lee, C.S. Enhanced mechanical compatibility of submicrocrystalline Ti-13Nb-13Zr alloy. *Mater. Sci. Eng. A* **2010**, *527*, 4914–4919. [CrossRef]

38. Matsumoto, H.; Yoshida, K.; Lee, S.-H.; Ono, Y.; Chiba, A. Ti-6Al-4V alloy with an ultrafine-grained microstructure exhibiting low-temperature-high-strain-rate superplasticity. *Mater. Lett.* **2013**, *98*, 209–212. [CrossRef]

39. Song, H.-W.; Zhang, S.-H.; Cheng, M. Dynamic globularization kinetics during hot working of a two phase titanium alloy with a colony alpha microstructure. *J. Alloys Compd.* **2009**, *480*, 922–927. [CrossRef]

40. Maccagno, T.M.; Jonas, J.J.; Hodgson, P.D. Spreadsheet modelling of grain size evolution during rod rolling. *ISIJ Int.* **1996**, *36*, 720–728. [CrossRef]

41. Ha, T.K.; Chang, Y.W. An internal variable theory of structural superplasticity. *Acta Mater.* **1998**, *46*, 2741–2749. [CrossRef]

metals

MDPI

Article

High Temperature Flow Response Modeling of Ultra-Fine Grained Titanium

Seyed Vahid Sajadifar and Guney Guven Yapici *

Mechanical Engineering Department, Ozyegin University, Istanbul 34794, Turkey;
seyedvahid.sajjadifar@ozu.edu.tr
* Author to whom correspondence should be addressed; guven.yapici@ozyegin.edu.tr; Tel.: +90-216-564-9115;
Fax: +90-216-564-9057.

Academic Editor: Heinz Werner Höppel
Received: 14 May 2015; Accepted: 14 July 2015; Published: 22 July 2015

Abstract: This work presents the mechanical behavior modeling of commercial purity titanium subjected to severe plastic deformation (SPD) during post-SPD compression, at temperatures of 600-900 °C and at strain rates of 0.001-0.1 s^{-1}. The flow response of the ultra-fine grained microstructure is modeled using the modified Johnson-Cook model as a predictive tool, aiding high temperature forming applications. It was seen that the model was satisfactory at all deformation conditions except for the deformation temperature of 600 °C. In order to improve the predictive capability, the model was extended with a corrective term for predictions at temperatures below 700 °C. The accuracy of the model was displayed with reasonable agreement, resulting in error levels of less than 5% at all deformation temperatures.

Keywords: severe plastic deformation; titanium; equal channel angular extrusion/pressing; high temperature; ultra-fine grained; Johnson-Cook model

---

1. Introduction

The advent of next generation materials with ultra-fine grains and nanostructures has led to further improvement of mechanical properties. Titanium and its alloys typically find use in various industries, including aerospace, defense, biomedical, energy, and automotive. In this respect, commercial purity titanium (CP Ti) has also received considerable attention due to the opportunity presented by remarkable combinations of strength and ductility by severe plastic deformation (SPD).

As a promising severe plastic deformation (SPD) method, equal channel angular extrusion/pressing (ECAE/P) can generate ultra-fine grained (UFG) structures with high yield strength, and ultimate tensile strength and accompanied by moderate ductility in various studies. The influence of ECAE routes on the microstructure and mechanical properties of CP Ti was investigated [1]. It was found that processing routes are effective in dictating the resulting grain morphology and structural refinement. Another study focused on the critical parameters of ECAE processing as a practical method for grain refinement [2]. Fan *et al.* [3] processed titanium using two-step SPD, consisting of eight passes of ECAE and cold rolling at room and liquid nitrogen temperature. Subsequent cold rolling of ECAE processed specimens led to further refinement of the microstructure, leading to strength levels higher than those of Ti-6Al-4V [4]. A similar work on post-ECAE rolling was also demonstrated in Zhu *et al.* [5]. Furthermore, evolution of crystallographic texture and anisotropy of post-ECAE rolled Ti was another topic of study where the best mechanical characteristics providing high ultimate tensile strength and ductility were revealed in the rolled flow plane samples, strained perpendicular to the rolling direction [6]. Recently, a comparative hardness study was conducted on CP and UFG Ti, indicating 2.5 times higher values for the latter, along with a more uniform microstructure after SPD [7]. As an interesting contribution, Segal [8] analyzed the commercialization of the ECAE process and discussed the optimization of ECAE process by control of contact friction, the channel geometry, strain/strain rate, billet shape, and punch/tool pressures.

In addition to the investigation of processing-structure-property relations, effective use of UFG materials requires efforts into the modeling of the mechanical behavior. For instance, processing of UFG materials requires the estimation of parameters to be applied during post-SPD deformation, including, but not limited to, rolling or forging. Therefore, flow stress characterization and modeling of deformation mechanics under high-temperature conditions is a critical topic, but has rarely been investigated for UFG materials. For the case of UFG Ti, a majority of the studies concentrated on high-temperature mechanical behavior and microstructural characterizations. Hoseini *et al.* [9] investigated the annealing behavior of UFG Ti, where thermal stability up to 450 °C was demonstrated. Long *et al.* [10] probed the mechanical behavior of UFG Ti in a wide temperature range by demonstrating the compressive deformation response from −196 to 600 °C. Recently, Meredith and Khan [11,12] have studied the mechanical properties and texture evolution of severely deformed titanium, at strain rates from 0.0001 to 2000 s^{-1} and temperatures from −196 to 375 °C.

On the modeling front, most investigations considered high temperature flow behavior prediction of coarse-grained Ti. Since the material flow behavior during hot deformation is complex, hardening and softening mechanisms are affected by strain rate as well as temperature. Thus, numerous research efforts have focused on the influence of strain rate and temperature on the flow stress using various constitutive models. A case on commercial purity Ti was presented using an Arrhenius type equation to propose constitutive equations at elevated temperatures up to 700 °C [13]. In another study on pure Ti, Zhang *et al.* [14] applied a model based on the Fields-Backofen equation to predict the tensile behavior up to 600 °C. A similar approach was followed to model the flow stress behavior of commercial pure titanium sheet at temperatures ranging from 350 to 500 °C [15]. Furthermore, neural network prediction of flow stress of a titanium alloy was investigated by Ping *et al.* [16]. Hyperbolic sine function based on the unified viscoplasticity theory was also utilized to model the flow behavior of a beta titanium alloy during hot deformation [17]. Another approach of simulating plastic deformation is proposed by Johnson and Cook [18] and has been utilized for modeling titanium alloys at elevated temperatures [19].

In spite of several studies on UFG Ti, constitutive modeling for describing the hot deformation characteristics of this material needs further investigation. Apart from recent attempts by Sajadifar and Yapici [20–22] considering the flow stress prediction of Ti subjected to eight passes ECAE, modeling of severely deformed materials in the high temperature regime has been a less frequent topic. In the present study, isothermal compression of ECAE processed CP titanium has been carried out at various temperatures and strain rates to characterize the flow behavior during hot deformation. The material behavior at high temperatures is predicted by introducing a modified Johnson-Cook model.

2. Materials and Experimental Procedures

The as-received commercial purity grade 2 Ti was received in bar form, with the chemical composition of 0.008% C, 0.041% Fe, 0.002% H, 0.006% N, 0.150% O and balance Ti represented in weight percent. The bars were coated with a graphite base lubricant before extrusion and were heated in a furnace to the deformation temperature of 300 °C where they were held for 1 h before extrusion. Finally, they were transferred to the 25.4 mm × 25.4 mm cross section, 90° angle ECAE die, which was preheated to 300 °C. Extrusion took place at a rate of 1.27 mm/s. Eight ECAE passes were performed following route E, accumulating a total strain of 9.24 in the as-processed material [23]. Route E was selected as the ECAE processing route, resulting in the largest fully worked region in a given billet with a high volume fraction of high-angle grain boundaries [24]. Route E consists of an alternating rotation of the billet by +180° and +90°, around its long axis, between successive passes. Following each extrusion pass, the billets were water quenched to maintain the microstructure achieved during ECAE. Lowest possible processing temperatures were crucial in preventing possible recrystallization and partly achieved using the sliding walls concept [8].

In exploring the hot deformation behavior of ECAE processed pure Ti, hot compression experiments were performed. The compression specimens were electro-discharge machined (EDM) in a rectangular block shape, 4 mm × 4 mm × 8 mm, with compression axis parallel to the extrusion direction. All samples were ground and polished to remove major scratches and eliminate the influence of a residual layer from EDM. The compression tests were conducted under isothermal conditions at three different strain rates of 0.001, 0.01, and 0.1 s^{-1} and at temperatures of 600, 700, 800, and 900 °C. All specimens were heated up to the deformation temperature and then the samples were deformed in a single loading step. The reduction in height at the end of the compression tests was 60% (true strain: 0.9) to avoid barreling and to capture the effect of both dynamic recovery and recrystallization on deformation behavior. In addition, lubrication with graphite was used during hot compression tests to decrease friction effects and minimize barreling [25]. The mechanical experiments were conducted inside a temperature-controlled furnace mounted on an Instron mechanical testing frame with uniformly heated samples having constant temperature profiles throughout the tests.

3. Results and Discussion

3.1. Effects of Temperature and Deformation Rate on the Microstructural Evolution

Increase of deformation temperature and decrease of deformation rate caused the growth of recrystallized grains in severely deformed titanium, as represented in Figure 1. Effect of deformation temperature on grain growth was observed to be more remarkable than that of deformation rate. At 900 °C, there is significant energy to promote nucleation and rapid growth of dynamically recrystallized grains. Deforming samples at 600 °C and 900 °C for longer periods led to a coarser structure since dynamically recrystallized grains had enough time to grow after nucleation.

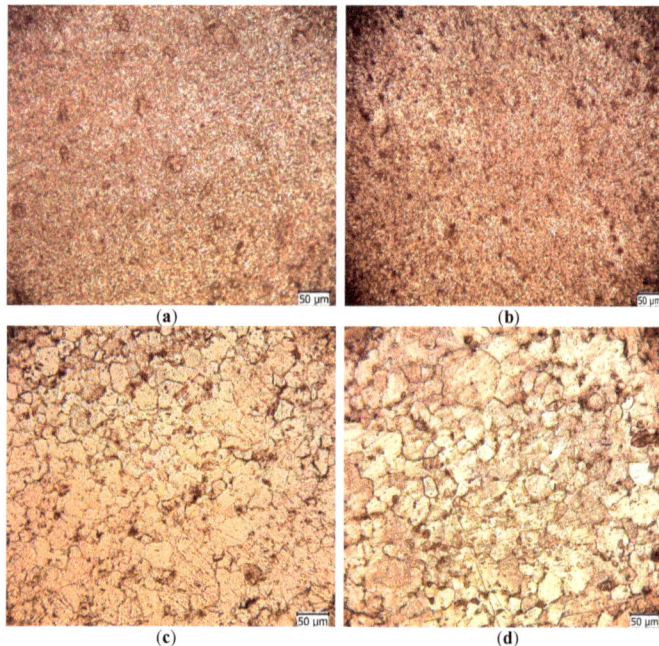

Figure 1. Microstructure of severely deformed titanium followed by hot compression test at (a) 600 °C-0.1s^{-1}, (b) 600 °C-0.001s^{-1}, (c) 900 °C-0.1s^{-1}, (d) 900 °C-0.001s^{-1}.

3.2. Constitutive Equation for the Modified Johnson-Cook Model

Flow stress during hot deformation is dependent on strain (ε), strain rate ($\dot{\varepsilon}$), and temperature (T), which can be described by the following function:

$$\sigma = f\left(\varepsilon, \dot{\varepsilon}, T\right) \tag{1}$$

The Johnson-Cook (JC) and its modified models are utilized in different materials to predict their flow stress behavior at elevated temperature [18]. The modified JC equation for hot compression is as follows [26]:

$$\sigma = \left(\sigma_P + B(T)\varepsilon^{n(T)}\right)\left[1 + C(\varepsilon,T)\ln\left(\frac{\dot{\varepsilon}}{\dot{\varepsilon}_0}\right)\right] \tag{2}$$

$B(T)$ and $n(T)$ are material constants, which are dependent on deformation temperature. $C(\varepsilon,T)$ is the material constant that is a function of both deformation temperature and strain. $\dot{\varepsilon}_0$ is the reference strain rate. The lowest strain rate (0.001 s^{-1}) is selected as the reference strain rate [26].

3.3. Calculation of B(T), n(T) and C(ε,T)

For the strain rate ($\dot{\varepsilon}$) of 0.001 s^{-1}, Equation (2) can be written as:

$$\sigma = \left(\sigma_P + B(T)\varepsilon^{n(T)}\right) \tag{3}$$

where σ_P is the peak stress. Taking the logarithm of Equation (3), gives:

$$\ln(\sigma_P - \sigma) = \ln[-B(T)] + n(T)\ln(\varepsilon) \tag{4}$$

The values of n and B can be obtained from Equation (4), according to the slope and intercept of the lines (Figure 2), respectively. The linear relationship was seen to be better at higher deformation temperatures in comparison with that of lower temperatures. This could be attributed to the stability of the UFG structure at lower deformation temperatures bringing higher discrepancy at 600 °C and 700 °C. The values of n and B are presented in Table 1.

Table 1. The values of B and n under different temperatures and at a strain rate of 0.001 s^{-1}.

Material Constants	600 °C	700 °C	800 °C	900 °C
$B(T)$, MPa	−60.099	−11.806	−1.696	−1.995
$n(T)$	4.311	3.251	1.319	0.123

With the substitution of $n(T)$ and $B(T)$ into Equation (2), respective values of $C(\varepsilon,T)$ can be calculated under various deformation conditions. The values of $C(\varepsilon,T)$ *versus* strain for different deformation conditions are obtained by linear fitting (Figure 3). As presented in Table 2 for selected true strain levels. For the sake of brevity, figures demonstrating the relation for the other deformation temperatures are not presented here.

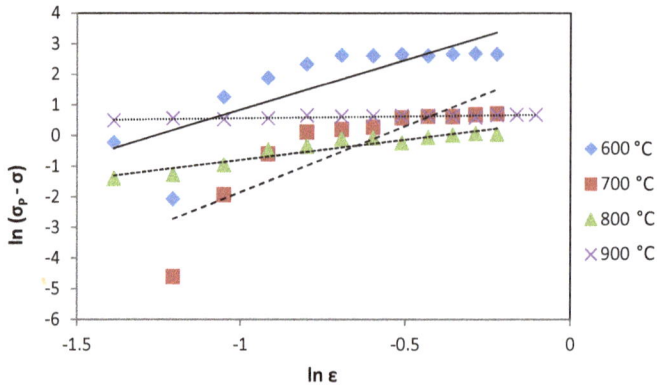

Figure 2. Relationship between $\ln(\sigma_P - \sigma)$ and $\ln(\varepsilon)$ at various temperatures and at a strain rate of $0.001 \ \text{s}^{-1}$.

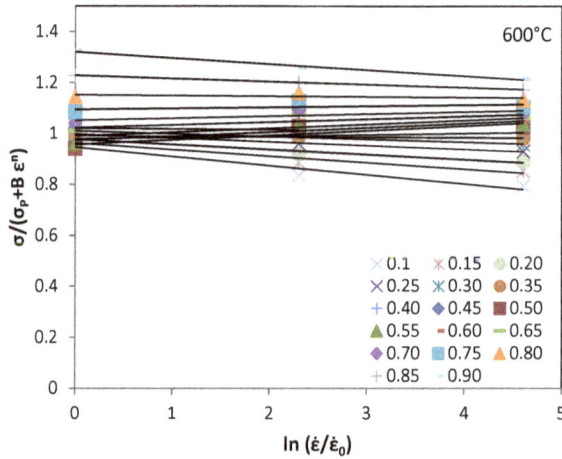

Figure 3. Relationship between $\dfrac{\sigma}{[\sigma_P + B(T)\varepsilon^{n(T)}]}$ and $\ln\left(\dfrac{\dot{\varepsilon}}{\dot{\varepsilon}_0}\right)$.

Table 2. The values of C at different deformation strain levels and temperatures.

Strain	600 °C	700 °C	800 °C	900 °C
0.20	−0.0201	−0.013	0.0005	−0.013
0.40	0.0055	−0.0005	0.0016	−0.0005
0.60	0.015	−0.0162	−0.0019	−0.0162
0.80	−0.0029	−0.0265	−0.0035	−0.0265

3.4. Prediction of Flow Stress by the Modified Johnson-Cook Modeling

The flow stress behavior of ECAE processed pure Ti can be predicted by applying $B(T)$, $n(T)$ and $C(\varepsilon,T)$ to Equation (2), as demonstrated in Figure 4. This model is precise in predicting the flow stress at all deformation conditions except at a temperature of 600 °C. At lower deformation temperatures and higher deformation rates, the severely deformed microstructure is partly retained, and as such, the model may lose its accuracy in determining the actual softening behavior.

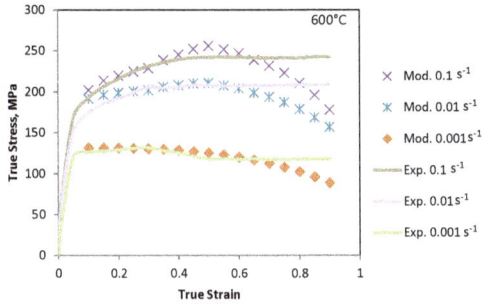

(a)

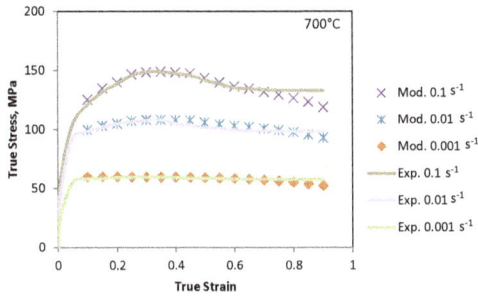

(b)

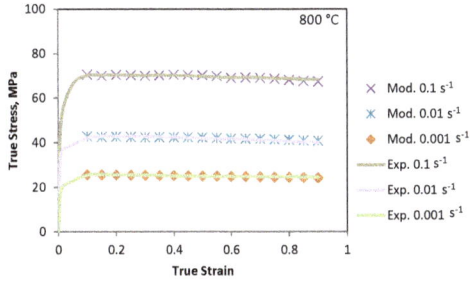

(c)

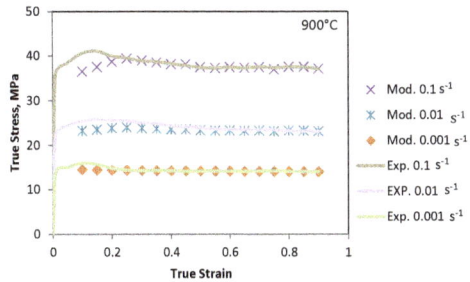

(d)

Figure 4. Comparison among model predictions and experimental results at: (**a**) 600 °C, (**b**) 700 °C, (**c**) 800 °C, (**d**) 900 °C.

Furthermore, it is worth noting that the difference between the simulated and experimental data is higher with increasing strain at 600 °C. Here, the modeled flow stress curves generally reach a peak and then decrease sharply, which is a pronounced characteristic of dynamic recrystallization. However, few experimental flow stress curves feature a peak followed by an immediate plateau. This is a well-known behavior of dynamic recovery and can be observed at the lowest deformation temperature.

3.5. Extension of the Johnson-Cook Model for Prediction of ECAE Processed Ti with a Corrective Term

When modeling the deformation at 600 °C, there is a considerable discrepancy between the experimental and model flow responses especially after the peak stress. The variation of the material constants, $n(T)$ and $B(T)$, as a function of temperature is plotted in Figure 5 exhibiting strong dependence on the deformation temperature. It is worth noting that $n(T)$ increases with decrease of deformation temperature. Moreover, it was seen that the absolute value of $B(T)$ shows a considerable increase with decrease of deformation temperature. However, at higher temperatures, $B(T)$ is fairly constant and can be considered to be independent of temperature. With the increase of deformation temperature, a similar behavior for $B(T)$ and $n(T)$ was reported for modeling the high temperature behavior of an aluminum alloy [26].

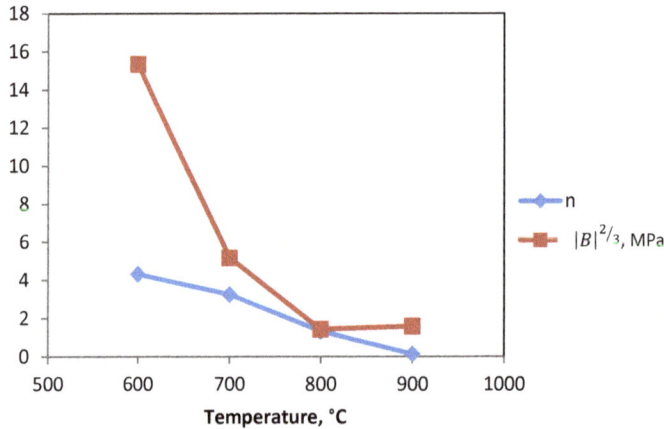

Figure 5. Deformation temperature *versus* n and $|B|^{2/3}$.

Another important factor affecting the predictions of this model is the amount of true strain. As can be observed in Figure 4a, agreement of the model with the experimental data gets worse from peak strain up to a true strain of 0.9. Considering that the error levels are comparably larger for deformation at 600 °C, a new term accounting for this behavior is added to Equation (2). This corrective term can be formulated as:

$$\text{Corrective Term} = \frac{(n(T) + \varepsilon)\left(\sigma_P \cdot \varepsilon^{n(T)}\right)}{|B(T)|^{\frac{2}{3}}} \qquad (5)$$

The corrective term depends on the degree of deformation temperature via the inclusion of the material constants, $n(T)$ and $B(T)$. With the dependence of corrective term on true strain, flow stress levels are expected to remain constant and a plateau behavior can be obtained. The modified flow response prediction at the lowest deformation temperature of 600 °C is shown in Figure 6 demonstrating better agreement with the experimental curve.

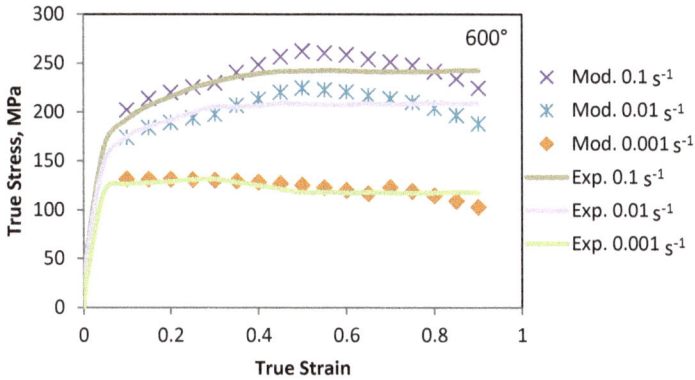

Figure 6. Comparison between improved model predictions and experimental results at 600 °C.

3.6. Verification of the Model

In order to verify the accuracy of the applied constitutive model, the deviations between the predicted stress (σ_P) and experimental stress (σ_E) values were obtained as:

$$E(\%) = \left| \frac{\sigma_P - \sigma_E}{\sigma_E} \right| \times 100 \qquad (6)$$

Figure 7 demonstrates the mean error values for the modeling approach based on the modified JC equation. It can be seen that the highest error level is less than 5% in all cases indicating reasonable agreement. In a separate study, models based on Arrhenius and dislocation density based formulations were employed for modeling the behavior of ECAE processed Ti in the warm to hot working regime. It can be seen that the error levels obtained using the current model is comparably less than those reported in [20,22].

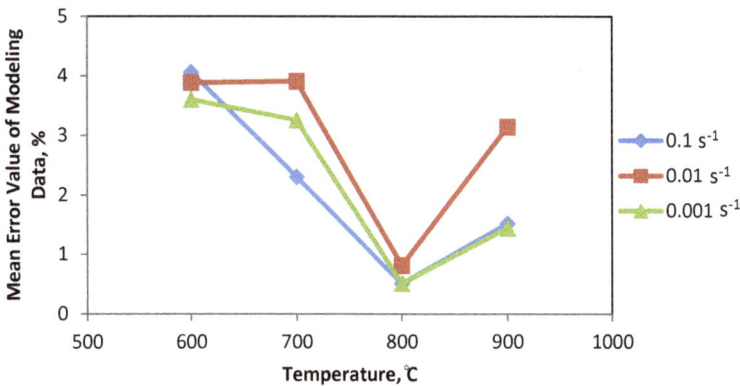

Figure 7. Mean error of the model for various deformation conditions.

Although both the modified JC and dislocation density based models exhibit decent predictive capability at relatively higher temperatures, their accuracy depends on how the active softening mechanisms are treated according to the deformation temperature. Another assumption of these constitutive models is that, they typically consider coarse grained microstructures. However, it was mentioned that the average grain size of UFG titanium remained less than 3.5 μm at 600 °C [9].

This finding was also confirmed in a different study [20,22]. On the other hand, the microstructure of ECAE processed Ti followed by hot compression tests (above 700 °C) contained grains with average size over 15 µm [22]. Therefore, the relatively fine grain size at lower temperatures could lead to the higher error levels observed in the prediction of the flow response at 600 °C.

4. Conclusions

The hot deformation behavior of ECAE processed pure Ti was studied by isothermal compression tests in temperatures ranging from 600 °C to 900 °C and strain rates ranging from 0.001 s^{-1} to 0.1 s^{-1}. The modified Johnson Cook model was utilized to predict the flow behavior of ECAE processed pure Ti leading to a satisfactory agreement with experimental data in all cases except 600 °C. The model is extended with a corrective term to improve the error levels at lower temperatures. This resulted in the highest error level to remain less than 5%, pointing to the probable use of the presented model in predicting the compressive response of ECAE processed Ti at elevated temperatures.

Acknowledgments: The authors would like to acknowledge the support from the European Union Seventh Framework Programme, Marie Curie Career Integration Grant, Project No: 304150, for conducting this investigation.

Author Contributions: G.G. Yapici designed the scope of the paper. S.V. Sajadifar prepared samples and conducted experiments. Both authors analyzed the results and participated in the modeling studies. S.V. Sajadifar wrote the initial draft and G.G. Yapici shaped the manuscript into its final form.

Conflicts of Interest: The authors declare no conflict of interest.

References

1. Stolyarov, V.V.; Zhu, Y.T.; Alexandrov, I.V.; Lowe, T.C.; Valiev, R.Z. Influence of ECAP routes on the microstructure and properties of pure Ti. *Mater. Sci. Eng. A* **2001**, *299*, 59–67. [CrossRef]
2. Valiev, R.Z.; Langdon, T.G. Principles of equal channel angular pressing as a processing tool for grain refinement. *Prog. Mater. Sci.* **2006**, *51*, 881–981. [CrossRef]
3. Fan, Z.; Jiang, H.; Sun, X.; Song, J.; Zhang, X.; Xie, C. Microstructures and mechanical deformation behaviors of ultrafine-grained commercial pure (grade 3) Ti processed by two-step severe plastic deformation. *Mater. Sci. Eng. A* **2009**, *527*, 45–51. [CrossRef]
4. Stolyarov, V.V.; Zhu, Y.T.; Alexandrov, I.V.; Lowe, T.C.; Valiev, R.Z. Grain refinement and properties of pure Ti processed by warm ECAP and cold rolling. *Mater. Sci. Eng. A* **2003**, *343*, 43–50. [CrossRef]
5. Zhu, Y.T.; Kolobov, Y.R.; Grabovetskaya, G.P.; Stolyarov, V.V.; Girsova, N.V.; Valiev, R.Z. Microstructures and mechanical properties of ultrafine-grained Ti foil processed by equal-channel angular pressing and cold rolling. *J. Mater. Res.* **2003**, *18*, 1011–1016. [CrossRef]
6. Yapici, G.G.; Karaman, I.; Maier, H.J. Mechanical flow anisotropy in severely deformed pure titanium. *Mater. Sci. Eng. A* **2006**, *434*, 294–302. [CrossRef]
7. Sanosh, K.P.; Balakrishnan, A.; Francis, L.; Kim, T.N. Vickers and Knoop Micro-hardness Behavior of Coarse-and Ultrafine-grained Titanium. *J. Mater. Sci. Technol.* **2010**, *26*, 904–907. [CrossRef]
8. Segal, V.M. Engineering and commercialization of equal channel angular extrusion (ECAE). *Mater. Sci. Eng. A* **2004**, *386*, 269–276. [CrossRef]
9. Hoseini, M.; Pourian, M.H.; Bridier, F.; Vali, H.; Szpunar, J.A.; Bocher, P. Thermal stability and annealing behaviour of ultrafine grained commercially pure titanium. *Mater. Sci. Eng. A* **2012**, *532*, 58–63. [CrossRef]
10. Long, F.W.; Jiang, Q.W.; Xiao, L.; Li, X.W. Compressive deformation behaviors of coarse- and ultrafine-grained pure titanium at different temperatures: a comparative study. *Mater. Trans.* **2011**, *52*, 1617–1622. [CrossRef]
11. Meredith, C.S.; Khan, A. The microstructural evolution and thermo-mechanical behavior of UFG Ti processed via equal channel angular pressing. *J. Mater. Process. Technol.* **2015**, *219*, 257–270. [CrossRef]
12. Meredith, C.S.; Khan, A. Texture evolution and anisotropy in the thermo-mechanical response of UFG Ti processed via equal channel angular pressing. *Int. J. Plasticity* **2012**, *30–31*, 202–217. [CrossRef]
13. Zeng, Z.; Jonsson, S.; Zhang, Y. Constitutive equations for pure titanium at elevated temperatures. *Mater. Sci. Eng. A* **2009**, *505*, 116–119. [CrossRef]

14. Zhang, Z.Y.; Yang, H.; Li, H.; Ren, N.; Wang, D. Quasi-static tensile behavior and constitutive modeling of large diameter thin-walled commercial pure titanium tube. *Mater. Sci. Eng. A* **2013**, *569*, 96–105. [CrossRef]

15. Tsao, L.C.; Wu, H.Y.; Leong, J.C.; Fang, C.J. Flow stress behavior of commercial pure titanium sheet during warm tensile deformation. *Mater. Des.* **2012**, *34*, 179–184. [CrossRef]

16. Ping, L.; Kemin, X.; Yan, L.; Jianrong, T. Neural network prediction of flow stress of Ti–15–3 alloy under hot compression. *J. Mater. Process. Technol.* **2004**, *148*, 235–238. [CrossRef]

17. Wang, Z.; Qiang, H.; Wang, X.; Wang, G. Constitutive model for a new kind of metastable beta titanium alloy during hot deformation. *Trans. Nonferrous Metals Soc. China* **2012**, *22*, 634–641. [CrossRef]

18. Johnson, G.R.; Cook, W.H. A constitutive model and data for metals subjected to large strains, high strain rates and high temperatures. In Proceedings of the 7th International Symposium on Ballistics, The Hague, The Netherlands, 19–21 April 1983; pp. 541–547.

19. Khan, A.S.; Suh, Y.S.; Kazmi, R. Quasi-static and dynamic loading responses and constitutive modeling of titanium alloys. *Int. J. Plasticity* **2004**, *20*, 2233–2248. [CrossRef]

20. Sajadifar, S.V.; Yapici, G.G. Workability characteristics and mechanical behavior modeling of severely deformed pure titanium at high temperatures. *Mater. Des.* **2014**, *53*, 749–757. [CrossRef]

21. Sajadifar, S.V.; Yapici, G.G. Hot Deformation Behavior of Ultra-fine Grained Pure Ti. *Adv. Mater. Res.* **2014**, *829*, 10–14. [CrossRef]

22. Sajadifar, S.V.; Yapici, G.G. Elevated temperature mechanical behavior of severely deformed pure titanium. *J. Mater. Eng. Perform.* **2014**, *23*, 1834–1844. [CrossRef]

23. Segal, V.M. Materials processing by simple shear. *Mater. Sci. Eng. A* **1995**, *197*, 157–164. [CrossRef]

24. Purcek, G.; Yapici, G.G.; Karaman, I.; Maier, H.J. Effect of commercial purity levels on the mechanical properties of ultrafine-grained titanium. *Mater. Sci. Eng. A* **2011**, *528*, 2303–2308. [CrossRef]

25. Li, L.X.; Peng, D.S.; Liu, J.A.; Liu, Z.Q. An experiment study of the lubrication behavior of graphite in hot compression tests of Ti-6Al-4V alloy. *J. Mater. Process. Technol.* **2001**, *112*, 1–5. [CrossRef]

26. Lin, Y.C.; Li, Q.F.; Xia, Y.C.; Li, L.T. A phenomenological constitutive model for high temperature flow stress prediction of Al-Cu-Mg alloy. *Mater. Sci. Eng. A* **2012**, *534*, 654–662. [CrossRef]

Chapter 3:
UFG Iron and Steels

metals

MDPI

Article

Development of Nanocrystalline 304L Stainless Steel by Large Strain Cold Working

Marina Odnobokova, Andrey Belyakov * and Rustam Kaibyshev

Belgorod State University, Pobeda 85, Belgorod 308015, Russia; odnobokova@bsu.edu.ru (M.O.); rustam_kaibyshev@bsu.edu.ru (R.K.)

* Author to whom correspondence should be addressed; belyakov@bsu.edu.ru; Tel.: +7-4722-585457; Fax: +7-4722-585417.

Academic Editor: Heinz Werner Höppel

Received: 29 March 2015; Accepted: 14 April 2015; Published: 22 April 2015

Abstract: The microstructural changes leading to nanocrystalline structure development and the respective tensile properties were studied in a 304L stainless steel subjected to large strain cold rolling at ambient temperature. The cold rolling was accompanied by the development of deformation twinning and martensitic transformation. The latter readily occurred at deformation microshear bands, leading the martensite fraction to approach 0.75 at a total strain of 3. The deformation twinning followed by microshear banding and martensitic transformation promoted the development of nanocrystalline structure consisting of a uniform mixture of austenite and martensite grains with their transverse sizes of 120–150 nm. The developed nanocrystallites were characterized by high dislocation density in their interiors of about 3×10^{15} m^{-2} and 2×10^{15} m^{-2} in austenite and martensite, respectively. The development of nanocrystalline structures with high internal stresses led to significant strengthening. The yield strength increased from 220 MPa in the original hot forged state to 1600 MPa after cold rolling to a strain of 3.

Keywords: austenitic stainless steel; severe plastic deformation; deformation twinning; strain-induced martensite; grain refinement; nanocrystalline structure; strengthening

1. Introduction

The large strain deformations are considered as promising methods for development of advanced structural steels and alloys with enhanced mechanical properties [1,2]. The significant improvement of mechanical properties of metallic materials subjected to severe plastic deformations is commonly attributed to the strain-induced ultrafine-grained or, even, nanocrystalline structures [3–5]. The ultrafine-grained materials have been shown to possess a unique combination of high strength and surprisingly large ductility [6]. The efficiency of cold working for processing the high-strength ultrafine-grained/nanocrystalline products depends remarkably on the kinetics of grain refinement during plastic deformation. Austenitic stainless steels are typical representative of metallic materials exhibiting rapid grain refinement upon cold working [7–10]. The grain refinement in these steels is promoted by an intensive grain subdivision, which is associated with deformation twinning followed by strain-induced martensitic transformation [9–13]. Therefore, the austenitic stainless steels can be easily produced in high-strength ultrafine-grained/nanocrystalline state by conventional cold working technique like plate rolling [10]. In spite of a number of research works dealing with nanocrystalline stainless steels processed by large strain cold working, however, the mechanisms of microstructure evolution, *i.e.*, a role of deformation twinning and strain-induced martensite, and their contribution to strengthening are still unclear.

The strengthening of metallic materials processed by large strain deformation is generally discussed in terms of either grain boundary strengthening [14] or dislocation strengthening [15,16].

The former is commonly evaluated as $\sigma_{GB} = K_\varepsilon D^{-0.5}$, where D is the grain size and K_ε is a constant; and the latter is related to a square root of dislocation density as $\sigma_{DISL} = \alpha Gb\rho^{0.5}$, where α, G, and b are a constant, the shear modulus, and the Burgers vector, respectively. Assuming that the grain boundary strengthening and the dislocation strengthening contribute independently to overall strength, a modified Hall-Petch-type relationship has been recently introduced to relate the yield strength of ultrafine-grained/nanocrystalline materials processed by severe plastic deformation to their microstructural parameters, *i.e.*, the grain size and dislocation density, in the following form [17–19]:

$$\sigma_{0.2} = \sigma_0 + K_\varepsilon D^{-0.5} + \alpha Gb\rho^{0.5} \tag{1}$$

Here, σ_0 is the strength of dislocation-free single crystal. Recent studies on severely deformed quasi-single phase ultrafine-grained/nanocrystalline materials have shown that the contribution from dislocation strengthening exceeds remarkably that from grain boundaries [20,21]. However, the strengthening mechanisms for ultrafine-grained/nanocrystalline materials such as metastable austenitic stainless steels, which experience martensitic phase transformation during cold working, have not been studied.

The aim of this study is to clarify the microstructural operating mechanisms, which are mainly responsible to the development of nanocrystalline structure in a typical chromium-nickel stainless steel during large strain cold rolling, and to investigate the strengthening mechanisms of the cold rolled steel, namely, the relationship between the microstructural parameters and strength contributions.

2. Experimental Section

A 304L-type austenitic steel (Fe-0.05%C-18.2%Cr-8.8%Ni-1.65%Mn-0.43%Si-0.05%P-0.04%S, all in wt%) with an initial grain size of 21 μm was hot forged and annealed at 1100 °C followed by air cooling. The plate rolling was carried out on samples with an initial cross section of 30×30 mm^2 at room temperature to various total true strains up to 3. In order to clarify the conventional Hall-Petch relationship for the present steel, several rolled samples were annealed to various recrystallized grain sizes at temperatures of 900–1150 °C. The strain hardening was studied by Vickers hardness tests with a load of 3 N. The microstructural characterization was performed using a JEM-2100 transmission electron microscope (TEM, JEOL Ltd., Tokyo, Japan) and a Nova Nanosem 450 scanning electron microscope equipped with an electron back-scatter diffraction (EBSD) analyzer (FEI, Hillsboro, OR, USA) on the sample sections normal to the transverse direction. The volume fractions of the ferrite were averaged through X-ray analysis, magnetic induction method and EBSD technique. The transverse grain size was measured on EBSD micrographs by a linear intercept method along the normal direction. The dislocation density was estimated by counting individual dislocations revealed by TEM in the grain/subgrain interiors. The tensile tests were carried out at room temperature by using specimens with a gage length of 6 mm and width of 3 mm. The equilibrium phase content was calculated with ThermoCalc software using TCFE6 database (ThermoCalc Software, Stockholm, Sweden).

3. Results and Discussion

3.1. Strain Hardening and Phase Transformation

The effects of cold rolling on the hardness and strain-induced martensite fraction are shown in Figure 1. The hardness drastically increases from about 1360 MPa to 4000 MPa during cold rolling to a total strain of 0.5. Then, the rate of strain hardening gradually slows down leading to a progressive increase in the hardness to above 5200 MPa as the total strain increases to 3. In contrast to strain hardening, the fraction of strain-induced martensite almost linearly increases with strain in the strain range of $0 < \varepsilon < 2$. Upon further rolling, the kinetics of the phase transformation becomes sluggish leading the strain-induced martensite fraction to approach 0.75, which is close to its saturation value of about 0.85 as predicted by ThermoCalc. It should be noted in Figure 1 that there is no direct correlation between the strain hardening and the martensite transformation. In other words, the change in hardening rate during the rolling does not provide similar change in the martensite fraction. Remarkable increase in the martensite fraction from 0.2 to 0.65 occurs in the strain range from 0.5 to 2, while the hardness increase does not exceed 20%.

3.2. Microstructure Evolution

Typical deformation microstructures that developed during cold rolling to different total strains are shown in Figure 2. The deformation microstructures combined with the inverse pole figures for the normal direction (vertical in Figure 2) are shown in left-hand figures, whereas the right-hand figures represent the austenite/martensite phase distribution. An early deformation is accompanied by the frequent development of deformation twinning, which is typical feature of austenitic steels with low stacking fault energy [9,10,13,20], followed by the martensitic transformation. Further deformation results in a flattening of the original grains and the development of microshear bands. The latter ones serve as preferential nucleation sites for the strain-induced martensite [11,22,23], resulting in significant increase in the martensite fraction at intermediate strains of around 1 as mentioned before in Figure 1. The grain flattening and the microshearing result in the wavy microstructure at large rolling strains. This microstructure is mainly composed by the strain-induced martensite, since its fraction comprises 0.75 at a large strain of 3. Therefore, the largely strained microstructure consists of highly elongated wavy martensite grains interleaved with chains of fine austenite grains.

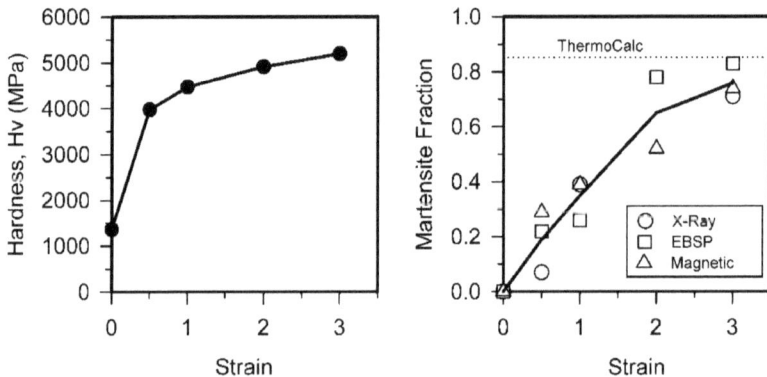

Figure 1. The effect of cold rolling strain on the hardness and strain-induced martensite fraction in a 304L stainless steel.

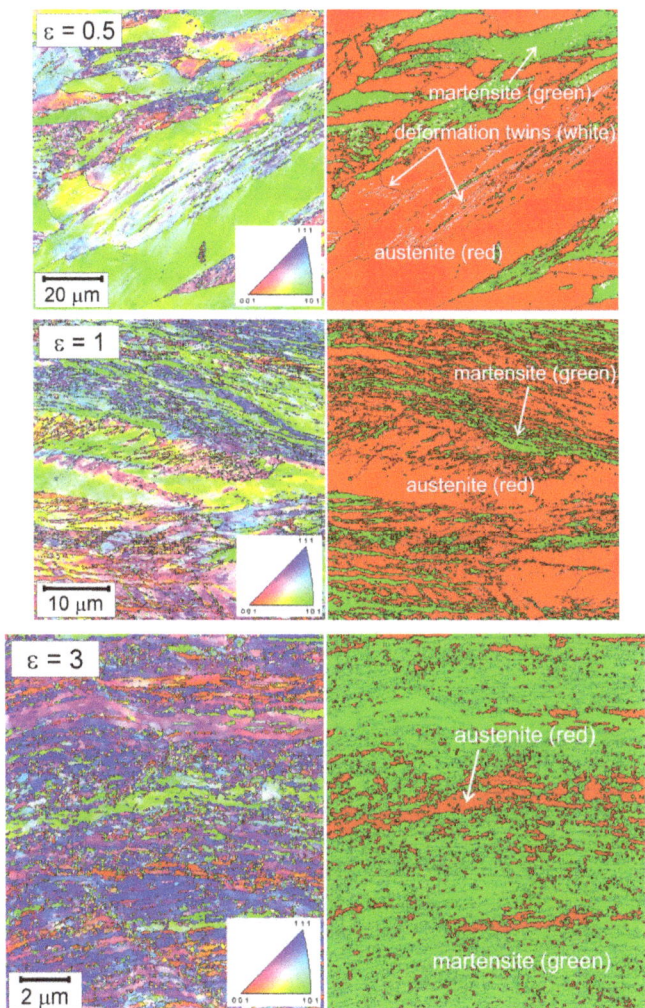

Figure 2. Deformation microstructures evolved in a 304L stainless steel during cold rolling various strains (ε). The black and white lines indicate the high-angle boundaries and twin boundaries, respectively. The inverse pole figures are shown for the normal direction.

Some details of the development of nanocrystalline structure in the present steel during cold rolling are shown in Figure 3. The microshear bands play an important role in the evolution of nanocrystalline structure. At intermediate strains, the strain-induced martensite readily develops at microshear bands. Hence, the microshear bands consist of alternating nanocrystallites of austenite and martensite (s. enlarged portion at $\varepsilon = 1$ in Figure 3). At large strains, the microshear bands cross over the flattened martensite crystallites. It should be noted that the martensite nanocrystallites evolved at large total strains are subdivided by high-angle grain boundaries (s. enlarged portion at $\varepsilon = 3$ in Figure 3).

The mechanisms of microstructure evolution during cold rolling are clearly reflected on the grain/phase boundary misorientation distributions evolved at different strain levels (Figure 4). The grain boundary misorientation distribution evolved at low to moderate stains of around 1 is characterized by three distinctive peaks against small angles below $10°$, large angles around $45°$, and large angles about $60°$. The first of them is clearly associated with a number of low-angle deformation subboundaries that are commonly brought out by plastic deformation [2]. The second one around $45°$ results from martensitic transformation. The orientation relationships between austenite and martensite in stainless steels are close to those predicted by Kurdjumov-Sachs and Nishiyama-Wasserman, which result in misorientations of $42.9°$ and $46°$, respectively, between austenite and martensite [9]. The third peak against $60°$ is, evidently, produced by deformation twinning, because the twin boundary misorientation in austenite is $60°$ around <111>. The misorientations of deformation subboundaries progressively increase during deformation [2]. Therefore, the fraction of low-angle subboundaries gradually decreases with increase in total strain. The pronounced deformation twinning at low to moderate strains seems to be exhausted at large strains. The corresponding $60°$ peak disappears at large strains. On the other hand, the strain-induced martensite continuously develops during the present cold rolling to a total strain of 3. It should be noted that grain boundaries in largely strained metals and alloys tend to exhibit random misorientation [2,24–26]. Therefore, the boundary misorientation distribution evolved in the stainless steel at large rolling strains looks like random distribution, which is superimposed with two peaks against small angles (deformation subboundaries are continuously developed) and large angles of $45°$ (resulting from martensitic orientation relationship).

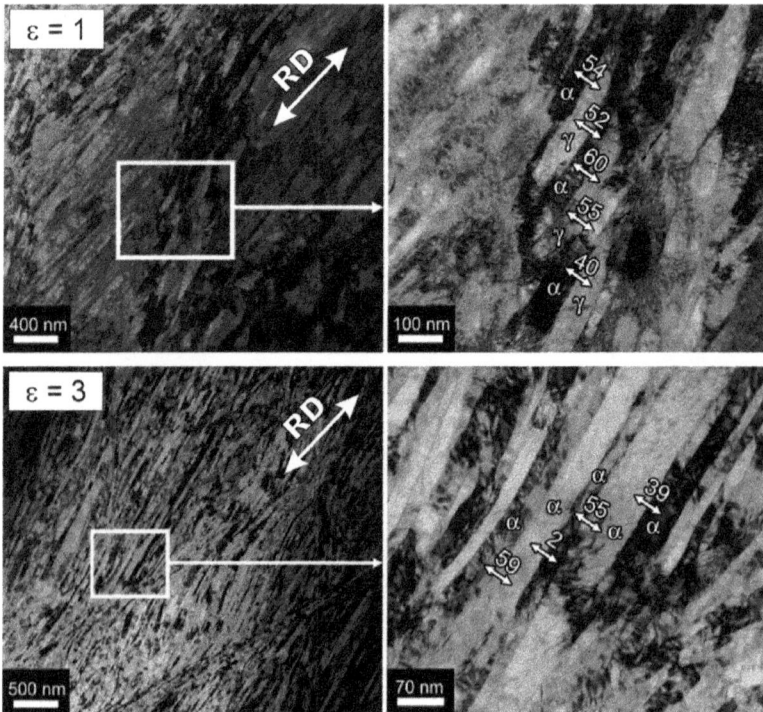

Figure 3. Fine structures evolved in a 304L stainless steel subjected to cold rolling to total strains of $\varepsilon = 1$ and $\varepsilon = 3$. The RD indicates the rolling direction. The numbers indicate the boundary misorientations in degrees.

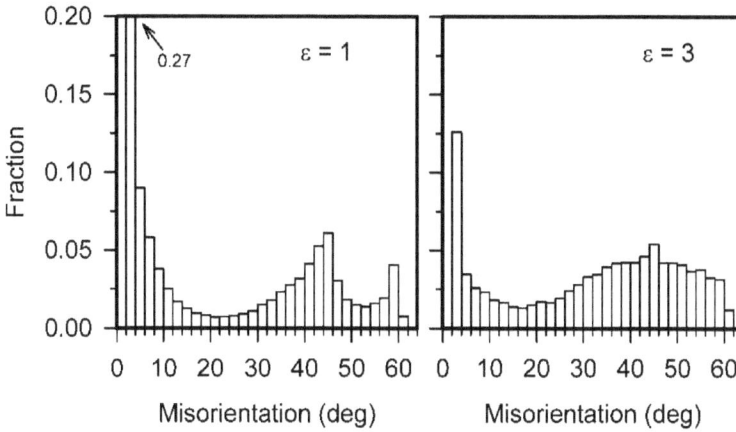

Figure 4. Grain boundary misorientation distributions evolved in a 304L stainless steel subjected to cold rolling to total strains of $\varepsilon = 1$ and $\varepsilon = 3$.

The strain effect on the transverse grain size and the dislocation density during cold rolling of the stainless steel (Figure 5) correlates with the strain hardening (Figure 1). The austenite grain size rapidly reduces to about 700 nm upon cold rolling to a strain of 1. Then, the strain effect on the grain refinement becomes less pronounced as strain increases. The transverse grain size of austenite gradually decreases to about 150 nm during cold rolling to a strain of 3. The transverse grain size of strain-induced martensite is characterized by similar strain dependence, although the martensite grains are finer than the austenite ones, especially, at relatively small strains. The martensite grain size finally attains about 120 nm at a total strain of 3. The dislocation density rapidly increases above 10^{15} m^{-2} at early deformation. Further cold rolling is accompanied by gradual increase in the dislocation density, which finally attains about 3×10^{15} m^{-2} in austenite and 2×10^{15} m^{-2} in martensite. A relatively low dislocation density in martensite may result from enhanced recovery in *bcc*-lattice.

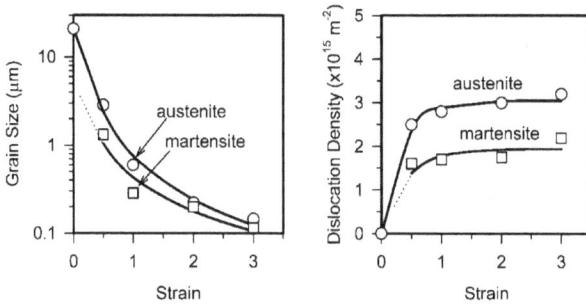

Figure 5. The effect of cold rolling strain on the transverse grain size and dislocation density in austenite and strain-induced martensite in a 304L stainless steel.

3.3. Tensile Behavior

The tensile stress-elongation curves for the 304L stainless steel subjected to cold rolling to different total strains are shown in Figure 6. The tensile behavior is characterized by a peak stress at relatively small strain followed by a decrease of the flow stress until fracture. The tensile strength increases, while the total elongation decreases with an increase in the previous rolling strain. The rolling to a strain of 3 results in significant increase in the yield strength from 220 MPa in the initial annealed state to 1600 MPa. The total elongation decreases correspondingly from 100% to 4%. Some microstructural parameters and mechanical properties of the steel samples cold rolled to different strains are listed in Table 1.

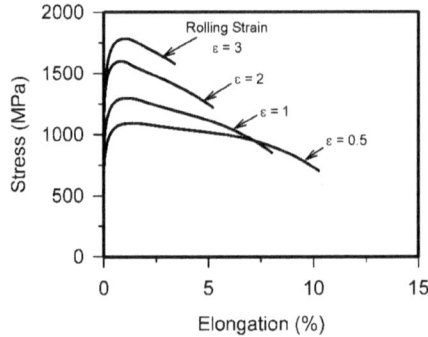

Figure 6. Engineering stress *vs.* plastic elongation curves for a 304L stainless steel subjected to cold rolling.

Table 1. The strain-induced martensite fraction (F_M), the grain size (D), the dislocation density (ρ), the yield strength ($\sigma_{0.2}$), the ultimate tensile strength (UTS), and the total elongation (δ) of a 304L stainless steel subjected to cold rolling to different strains (ε). The indexes of A and M indicate the austenite and martensite, respectively.

ε	F_M	D_A, μm	D_M, μm	ρ_A, 10^{14} m^{-2}	ρ_M, 10^{14} m^{-2}	$\sigma_{0.2}$, MPa	UTS, MPa	δ, %
0	0	21	-	0.02	-	220	600	100
0.5	0.20	2.85	1.33	25	16	950	1090	10
1	0.35	0.6	0.29	28	17	1160	1295	8
2	0.65	0.22	0.2	30	18	1485	1600	5
3	0.75	0.145	0.115	32	22	1595	1785	4

3.4. Strengthening Mechanisms

The yield strength of the present 304L steel subjected to large strain cold rolling can be expressed by the modified Hall-Petch relationship (Equation (1)), taking into account separate contributions of austenite and martensite to overall strength:

$$\sigma_{0.2} = F_A(\sigma_{0A} + K_{\varepsilon A}D_A^{-0.5} + \alpha_A G_A b_A \rho_A^{0.5}) + F_M(\sigma_{0M} + K_{\varepsilon M}D_M^{-0.5} + \alpha_M G_M b_M \rho_M^{0.5}) \quad (2)$$

where indexes of A and M indicate austenite and martensite, respectively, F_A and F_M are the austenite and martensite fractions, *i.e.*, $F_A + F_M = 1$. The values of σ_0 and K_ε can be obtained from conventional Hall-Petch relationship. To clarify the Hall-Petch relationship for the present steel, several annealed samples with statically recrystallized microstructures were subjected to tensile tests. The corresponding relationship between the austenite grain size and the yield strength is shown in Figure 7. Note here that the grain sizes were evaluated as the mean grain boundary spacing, counting all high-angle boundaries including twin boundaries. It is clearly seen in Figure 7 that the yield strength can be related to the austenite grain size as follows:

$$\sigma_{0.2A} = 180 + 240D_A^{-0.5} \quad (3)$$

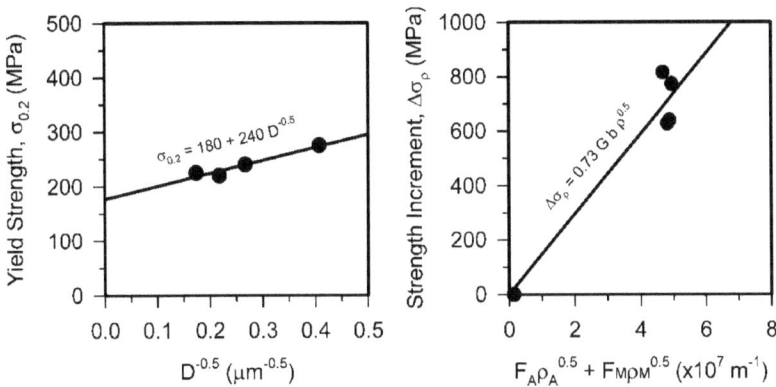

Figure 7. Hall-Petch relationship for a 304L stainless steel with annealed recrystallized microstructure and the effect of dislocation density (ρ) on the strength increment ($\Delta\sigma_\rho$).

The σ_{0A} = 180 MPa is quite close to those about 200 MPa reported for various austenitic steels [9,19,27]. The grain size strengthening in martensite can be evaluated by the following relationship, which has been obtained for recrystallized Fe–15%Cr steel [28]:

$$\sigma_{0.2M} = 180 + 240D_M^{-0.5} \tag{4}$$

Then, the dislocation strengthening can be estimated using the obtained data. The strength increment associated with dislocation strengthening from Equation (2) reads:

$$\Delta\sigma_\rho = \sigma_{0.2} - F_A\sigma_{0.2A} - F_M\sigma_{0.2M} = F_A\alpha_A G_A b_A \rho_A^{0.5} + F_M\alpha_M G_M b_M \rho_M^{0.5} \tag{5}$$

Note here that the shear modulus, G = 81 GPa, and Burgers vector, b = 0.25 nm, are almost the same for austenite and martensite (ferrite) [29]. Assuming that $\alpha_A = \alpha_M$, Equation (5) can be simplified as follows:

$$\Delta\sigma_\rho = \alpha Gb \left(F_A\rho_A^{0.5} + F_M\rho_M^{0.5}\right) \tag{6}$$

Figure 7 shows the relationship between the dislocation density and the corresponding strength increment. Thus, the value of α = 0.73 is obtained from Figure 7. It should be noted that almost the same values of α have been used for calculation of dislocation strengthening in various alloys [19–21,30–32]. Finally, the following expression for the yield strength of the present steel subjected to cold rolling can be obtained:

$$\sigma_{0.2} = F_A 180 + F_M 120 + 240(F_A D_A^{-0.5} + F_M D_M^{-0.5}) + 0.73Gb(F_A\rho_A^{0.5} + F_M\rho_M^{0.5}) \tag{7}$$

The relationship between the experimental yield stress and calculated by Equation (7) is shown in Figure 8. It is clearly seen that the yield strengths obtained by the modified Hall-Petch type equation are quite coincident with the experimental results. Figure 8 also shows the contributions of different strengthening mechanisms, *i.e.*, the austenite dislocation density ($\Delta\sigma_{\rho A}$), the austenite grain size ($\Delta\sigma_{DA}$), the martensite dislocation density ($\Delta\sigma_{\rho M}$), and the martensite grain size ($\Delta\sigma_{DA}$), into overall strength, taking into account the change in the austenite/martensite fraction during cold rolling. At small to moderate strains, the strengthening of the cold worked austenitic stainless steel is mainly provided by drastic increase in the dislocation density in the austenite. The strength contribution from austenite grain size becomes comparable with that from austenite dislocation density at rather large strains, when the grain size decreases to nano-scale level. The dislocation density and grain size of strain-induced martensite contribute to overall strength in similar manner as the austenite does.

Namely, the strength increment from dislocation density significantly exceeds that from grain size at relatively small strains, whereas the strength increments from dislocation density and grain size become almost the same at large strains. However, the difference between the strength increments from dislocation density and grain size in the strain-induced martensite is much less pronounced than that in the austenite in the range of moderate to large strains. After rolling to a large total strain of 3, the same strengthening from martensite dislocation density and martensite grain size is observed (Figure 8).

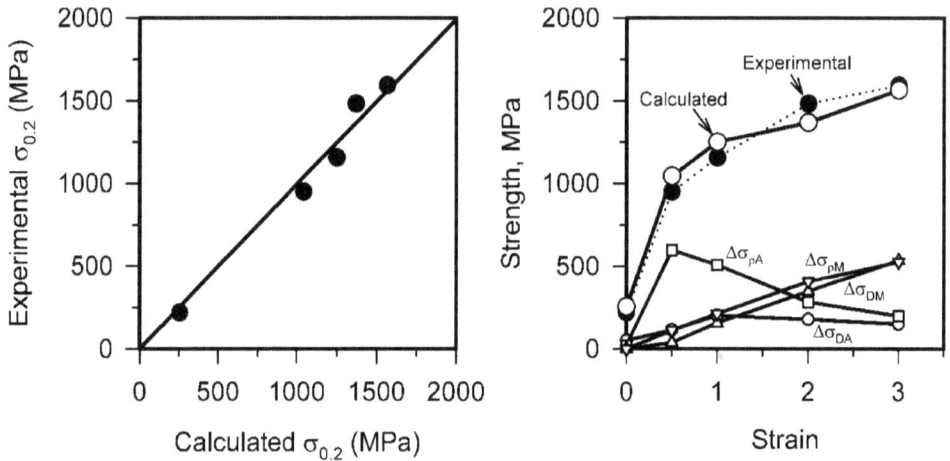

Figure 8. The relationship between experimental and calculated yield strength ($\sigma_{0.2}$) and the contribution of grain size strengthening ($\Delta\sigma_D$) and dislocation strengthening ($\Delta\sigma_\rho$) into overall strength on a 304L stainless steel subjected to cold rolling. The indexes of A and M indicate the austenite and martensite, respectively.

4. Conclusions

The microstructural evolution and corresponding mechanical properties of a 304L austenitic stainless steel subjected to large strain cold rolling at room temperature were studied. The main results can be summarized as follows:

1. The cold rolling was accompanied by a rapid increase in the dislocation density, which exceed 10^{15} m^{-2} after straining to 0.5. Features of microstructural changes in the austenitic stainless steel during cold deformation were the deformation twinning and the development of strain-induced martensitic transformation, which resulted in martensite fraction of 0.75 after rolling to a strain of 3. Both the deformation twinning and strain-induced martensite led to the rapid grain refinement. The nanocrystalline structure consisting of austenite and martensite grains with transverse grain sizes of 145 nm and 115 nm, respectively, was developed at a large total strain of 3.

2. The development of nanocrystalline structure provided significant strengthening. The yield strength increased from about 950 MPa to 1600 MPa with an increase in the total strain from 0.5 to 3. Considering the dislocation density (ρ) and grain size (D) as main contributors to overall strengthening, the following relationship for yield strength was obtained:

$$\sigma_{0.2} = F_A 180 + F_M 120 + 240(F_A D_A^{-0.5} + F_M D_M^{-0.5}) + 0.73Gb(F_A \rho_A^{0.5} + F_M \rho_M^{0.5})$$

where F_A and F_M are volume fractions of austenite and martensite, respectively, G is the shear modulus, b is Burgers vector, the indexes of A and M indicate austenite and martensite, respectively. The obtained results suggested that the strength increment from dislocation density remarkably exceeds that from grain size at small to moderate strains, whereas this difference gradually decreases during subsequent deformation to large total strains.

Acknowledgments: The authors gratefully acknowledge the financial support from the Ministry of Education and Science, Russia, (Belgorod State University project No. 1683). The authors are grateful to the personnel of the Joint Research Centre, Belgorod State University, for their assistance with the instrumental analysis.

Author Contributions: The present paper is a result of fruitful collaboration of all co-authors. R. Kaibyshev designed the research theme; A. Belyakov selected the experimental procedure; M. Odnobokova carried out the experimental study. Then, all co-authors discussed and analyzed the obtained results.

Conflicts of Interest: The authors declare no conflict of interest.

References

1. Valiev, R.Z.; Islamgaliev, R.K.; Alexandrov, I.V. Bulk nanostructured materials from severe plastic deformation. *Prog. Mater. Sci.* **2000**, *45*, 103–189. [CrossRef]
2. Sakai, T.; Belyakov, A.; Kaibyshev, R.; Miura, H.; Jonas, J.J. Dynamic and post-dynamic recrystallization under hot, cold and severe plastic deformation conditions. *Prog. Mater. Sci.* **2014**, *60*, 130–207. [CrossRef]
3. Valiev, R.Z.; Alexandrov, I.V.; Zhu, Y.T.; Lowe, T.C. Paradox of strength and ductility in metals processed by severe plastic deformation. *J. Mater. Res.* **2002**, *17*, 5–8. [CrossRef]
4. Kimura, Y.; Inoue, T.; Yin, F.; Tsuzaki, K. Inverse temperature dependence of toughness in an ultrafine grain-structure steel. *Science* **2008**, *320*, 1057–1060. [CrossRef] [PubMed]
5. Estrin, Y.; Vinogradov, A. Extreme grain refinement by severe plastic deformation: A wealth of challenging science. *Acta Mater.* **2013**, *61*, 782–817. [CrossRef]
6. Stolyarov, V.V.; Valiev, R.Z.; Zhu, Y.T. Enhanced low-temperature impact toughness of nanostructured Ti. *Appl. Phys. Lett.* **2006**, *88*, 041905. [CrossRef]
7. Misra, R.D.K.; Zhang, Z.; Venkatasurya, P.K.C.; Somani, M.C.; Karjalainen, L.P. Martensite shear phase reversion-induced nanograined/ultrafine-grained Fe–16Cr–10Ni alloy: The effect of interstitial alloying elements and degree of austenite stability on phase reversion. *Mater. Sci. Eng. A* **2010**, *527*, 7779–7792. [CrossRef]
8. Rezaee, A.; Kermanpur, A.; Najafizadeh, A.; Moallemi, M. Production of nano/ultrafine grained AISI 201L stainless steel through advanced thermo-mechanical treatment. *Mater. Sci. Eng. A* **2011**, *528*, 5025–5029. [CrossRef]
9. Shakhova, I.; Dudko, V.; Belyakov, A.; Tsuzaki, K.; Kaibyshev, R. Effect of large strain cold rolling and subsequent annealing on microstructure and mechanical properties of an austenitic stainless steel. *Mater. Sci. Eng. A* **2012**, *545*, 176–186. [CrossRef]
10. Belyakov, A.; Odnobokova, M.; Kipelova, A.; Tsuzaki, K.; Kaibyshev, R. Nanocrystalline structures and tensile properties of stainless steels processed by severe plastic deformation. *IOP Conf. Ser. Mater. Sci. Eng.* **2014**, *63*, 012156. [CrossRef]
11. Olson, G.B.; Cohen, M. Kinetics of strain-induced martensitic nucleation. *Metall. Trans. A* **1975**, *6*, 791–795. [CrossRef]
12. Nakada, N.; Ito, H.; Matsuoka, Y.; Tsuchiyama, T.; Takaki, S. Deformation-induced martensitic transformation behavior in cold-rolled and cold-drawn type 316 stainless steels. *Acta Mater.* **2010**, *58*, 895–903. [CrossRef]
13. Lee, T.-H.; Shin, E.; Oh, C.-S.; Ha, H.-Y.; Kim, S.-J. Correlation between stacking fault energy and deformation microstructure in high-interstitial-alloyed austenitic steels. *Acta Mater.* **2010**, *58*, 3173–3186. [CrossRef]
14. Armstrong, R.; Codd, I.; Douthwaite, R.M.; Petch, N.J. The plastic deformation of polycrystalline aggregates. *Philos. Mag.* **1962**, *7*, 45–58. [CrossRef]
15. Mecking, H.; Kocks, U.F. Kinetics of flow and strain-hardening. *Acta Metall.* **1981**, *29*, 1865–1875. [CrossRef]
16. Estrin, Y.; Toth, L.S.; Molinari, A.; Brechet, Y. A dislocation-based model for all hardening stages in large strain deformation. *Acta Mater.* **1998**, *46*, 5509–5522. [CrossRef]
17. Hughes, D.A.; Hansen, N. Microstructure and strength of nickel at large strains. *Acta Mater.* **2000**, *48*, 2985–3004. [CrossRef]
18. Hansen, N. Hall-Petch relation and boundary strengthening. *Scr. Mater.* **2004**, *51*, 801–806. [CrossRef]
19. Yanushkevich, Z.; Mogucheva, A.; Tikhonova, M.; Belyakov, A.; Kaibyshev, R. Structural strengthening of an austenitic stainless steel subjected to warm-to-hot working. *Mater. Charact.* **2011**, *62*, 432–437. [CrossRef]

20. Kusakin, P.; Belyakov, A.; Haase, C.; Kaibyshev, R.; Molodov, D.A. Microstructure evolution and strengthening mechanisms of Fe–23Mn–0.3C–1.5Al TWIP steel during cold rolling. *Mater. Sci. Eng. A* **2014**, *617*, 52–60. [CrossRef]

21. Mishnev, R.; Shakhova, I.; Belyakov, A.; Kaibyshev, R. Deformation microstructures, strengthening mechanisms, and electrical conductivity in a Cu–Cr–Zr alloy. *Mater. Sci. Eng. A* **2015**, *629*, 29–40. [CrossRef]

22. Kipelova, A.; Odnobokova, M.; Belyakov, A.; Kaibyshev, R. Microstructure evolution in a 304-type austenitic stainless steel during multidirectional forging at ambient temperature. *Mater. Sci. Forum* **2014**, *783–786*, 831–836. [CrossRef]

23. Odnobokova, M.; Kipelova, A.; Belyakov, A.; Kaibyshev, R. Microstructure evolution in a 316L stainless steel subjected to multidirectional forging and unidirectional bar rolling. *IOP Conf. Ser. Mater. Sci. Eng.* **2014**, *63*, 012060. [CrossRef]

24. Mishin, O.V.; Gottstein, G. Microstructural aspects of rolling deformation in ultrafine-grained copper. *Philos. Mag. A* **1998**, *78*, 373–388. [CrossRef]

25. Mishin, O.V.; Bowen, J.R.; Lathabai, S. Quantification of microstructure refinement in aluminium deformed by equal channel angular extrusion: Route A vs. route Bc in a 90° die. *Scr. Mater.* **2010**, *63*, 20–23. [CrossRef]

26. Tikhonova, M.; Kuzminova, Y.; Fang, X.; Wang, W.; Kaibyshev, R.; Belyakov, A. Σ3 CSL boundary distributions in an austenitic stainless steel subjected to multidirectional forging followed by annealing. *Philos. Mag.* **2014**, *94*, 4181–4196. [CrossRef]

27. Young, C.M.; Sherby, O.D. Subgrain formation and subgrain-boundary strengthening in iron-based materials. *J. Iron Steel Inst.* **1973**, *211*, 640–647.

28. Belyakov, A.; Tsuzaki, K.; Kimura, Y.; Mishima, Y. Tensile behaviour of submicrocrystalline ferritic steel processed by large-strain deformation. *Philos. Mag. Lett.* **2009**, *89*, 201–212. [CrossRef]

29. Frost, H.J.; Ashby, M.F. *Deformation Mechanism Maps*; Pergamon Press: Oxford, UK, 1982.

30. Huang, X.; Morito, S.; Hansen, N.; Maki, T. Ultrafine structure and high strength in cold-rolled martensite. *Metall. Mater. Trans. A* **2012**, *43A*, 3517–3531. [CrossRef]

31. Harrell, T.J.; Topping, T.D.; Wen, H.; Hu, T.; Schoenung, J.M.; Lavernia, E.J. Microstructure and strengthening mechanisms in an ultrafine grained Al-Mg-Sc alloy produced by powder metallurgy. *Metall. Mater. Trans. A* **2014**, *45A*, 6329–6343. [CrossRef]

32. Khamsuk, S.; Park, N.; Gao, S.; Terada, D.; Adachi, H.; Tsuji, N. Mechanical properties of bulk ultrafine grained aluminum fabricated by torsion deformation at various temperatures and strain rates. *Mater. Trans.* **2014**, *55*, 106–113. [CrossRef]

metals

MDPI

Article

Ultrafine-Grained Austenitic Stainless Steels X4CrNi18-12 and X8CrMnNi19-6-3 Produced by Accumulative Roll Bonding

Mathis Ruppert *, Lisa Patricia Freund, Thomas Wenzl, Heinz Werner Höppel and Mathias Göken

Department of Materials Science and Engineering, Institute I: General Materials Properties,
Friedrich-Alexander Universität Erlangen-Nürnberg, Martensstr. 5, 91058 Erlangen, Germany;
lisa.freund@fau.de (L.P.F.); thomas.wenzl@fau.de (T.W.); hwe.hoeppel@fau.de (H.W.H.);
mathias.goeken@ww.uni-erlangen.de (M.G.)

* Author to whom correspondence should be addressed; mathias.ruppert@fau.de; Tel.: +49-9131-8527478;
Fax: +49-9131-8527504.

Academic Editor: Hugo F. Lopez
Received: 30 March 2015; Accepted: 4 May 2015; Published: 7 May 2015

Abstract: Austenitic stainless steels X4CrNi18-12 and X8CrMnNi19-6-3 were processed by accumulative roll bonding (ARB). Both materials show an extremely high yield strength of 1.25 GPa accompanied by a satisfactory elongation to failure of up to 14% and a positive strain rate sensitivity after two ARB cycles. The strain-hardening rate of the austenitic steels reveals a stabilization of the stress-strain behavior during tensile testing. Especially for X8CrMnNi19-6-3, which has an elevated manganese content of 6.7 wt.%, necking is prevented up to comparatively high plastic strains. Microstructural investigations showed that the microstructure is separated into ultrafine-grained channel like areas and relatively larger grains where pronounced nano-twinning and martensite formation is observed.

Keywords: accumulative roll bonding (ARB); austenitic steel; ultrafine-grained microstructure; strength; nano-twinning; strain rate sensitivity

---

1. Introduction

Accumulative roll bonding (ARB) [1] as a process of severe plastic deformation is one of the most effective methods for the production of bulk ultrafine-grained (UFG) materials with a median grain size smaller than 1 μm. The microstructural evolution during ARB is well described in literature, see for example [2–4] for details. The mechanical properties of those materials are frequently claimed to be favorable compared to their conventionally grained (CG) counterparts, as a good combination of high strength and satisfactory ductility can be achieved [5–7]. The enhanced ductility is often brought into connection with the enhanced strain rate sensitivity [3,6,8,9] of those materials. Moreover, strain rate sensitivity is strongly related to an increased fraction of high angle boundaries [4], as they can act as sources and sinks for dislocations [3]. Although there are numerous publications on ARB available, only a couple of them are dealing with ARB-processing of steel-sheets. Among those, the majority concerns interstitial free bcc steels (IF-steels) with a rather low content of alloying elements, see for example [1,10–12]. However, there is very little literature available about ARB of austenitic steels, which will be focused on in the following. Kitahara *et al.* [13] investigated the martensite transformation of an ultrafine-grained Fe-28.5at.-%Ni-alloy with single phase metastable austenite at room temperature. They performed the accumulative roll bonding process with sheets that were pre-heated at 500 °C for 600 s up to 5 cycles. However, each cycle was divided into two passes with a thickness reduction <50%. They achieved an ultrafine-grained microstructure with a mean grain size of 230 nm and a yield strength that was increased by a factor of 4.9 compared to the initial material. Moreover, they showed that the martensite transformation starting temperature decreases with the number of ARB cycles. Another study about ARB of austenitic steel was published by Jafarian *et al.* [14]. They investigated the microstructure and texture development in a Fe-24Ni-0.3C (wt.%) austenitic steel up to 6 ARB-cycles and subsequent annealing. The processing was performed at 600 °C. The texture was found to change from a copper orientation after 1 cycle towards a strong brass component after 6 cycles of ARB. Shen *et al.* [15] performed an accumulative cold rolling process with a thickness reduction of 17% each pass, using sheets of a commercial austenitic stainless steel 304SS, which was pre-heated up to 400 °C prior to each pass. Due to various subsequent heat-treatments, they could achieve materials with different grain sizes. The tensile samples reached a yield strength of up to 1.8 GPa, yet rather small elongation to failure of about 6%. Furthermore twinning and slip of partial dislocations were found to dominate plastic deformation in the ultrafine-grained state. Li *et al.* [16] processed sheets of an austenitic 36%Ni (mass-%) steel up to 6 cycles, with a pre-ARB heat-treatment at 500 °C. They found a rather small grain size of 150 nm and a high misorientation concerning the high angle boundaries.

As no information is available in literature, this work is focused on the strain rate sensitivity of ARB-processed ultrafine-grained austenitic stainless steels. Moreover, the effect of ultrafine grains and pronounced nano-twinning on the mechanical properties of commercially available and technically relevant alloys was addressed.

2. Experimental Section

Austenitic steels X4CrNi18-12 (1.4303) and X8CrMnNi19-6-3 (1.4376) were used as initial sheet materials for accumulative roll bonding. Due to the high Ni-content, the austenitic phase of X4CrNi18-12 is stabilized, which leads to a higher cold forming capability. This is desirable, as the material is severely strained during ARB-processing. The austenitic steel X8CrMnNi19-6-3 is a metastable one, which contains 6.7 wt.% of manganese. This induces twinning by plastic deformation, which might lead to satisfactory ductility. The material was delivered by Thyssen Krupp Nirosta GmbH and the chemical composition can be found in Table 1.

Table 1. Chemical compositions of the processed austenitic steels X4CrNi18-12 and X8CrMnNi19-6-3.

Alloy	Composition in wt.%								
	C	Si	Mn	Cr	Ti	P	S	Mo	Ni
X4CrNi18-12	0–0.06	-	0–2.0	17.0–19.0	-	-	-	-	11.0–13.0
X8CrMnNi19-6-3	0.025	0.46	6.76	17.43	0.001	0.029	0.0008	0.23	4.03

The sheets of X4CrNi18-12 and X8CrMnNi19-6-3 had an initial geometry of 25 × 150 × 1 mm (width × length × thickness) and were processed up to three and two cycles of ARB, respectively. Henceforth, the number of performed ARB cycles is denoted by N0-N3. Hereby one cycle of ARB equals a v. Mises equivalent strain of 0.8. During each cycle, the sheets were degreased in acetone and wire brushed with a rotating steel brush. Afterwards, the sheets were pre-heated in an electrical furnace for 5 min at 300 °C and finally roll bonded with a thickness reduction of 50%. After each cycle, edge cracking was cut off and the sheets were prepared accordingly to the scheme described above before the next cycle. In order to determine the mechanical properties of the processed sheets, both Vickers hardness measurements and tensile testing were performed. Therefore, a hardness measurement unit Leco V-100A and an Instron 4505 universal testing machine (Hegewald & Peschke MPT GmbH, Nossen, Germany) for uniaxial tensile testing were utilized. The hardness measurements were conducted at the sheet plane, the rolling plane and the transversal plane. Tensile testing was conducted in rolling direction at room temperature and at strain rates of $10^{-3}s^{-1}$, $10^{-4}s^{-1}$ and $10^{-5}s^{-1}$

in order to determine the strain rate sensitivity. Moreover, microstructural characterization was done using a Zeiss Cross Beam 1540 EsB (Carl Zeiss AG, Oberkochen, Germany) scanning electron microscope in backscattered electron contrast at an acceleration voltage of 11 kV and a working distance of 7–8 mm, as well as a Philips CM 200 transmission electron microscope operated at 200 kV (FEI, Hillsboro, OR, USA).

3. Results and Discussion

In Figure 1, the results of the hardness measurements in sheet plane, rolling plane and transversal plane are shown for X4CrNi18-12 and for X8CrMnNi19-6-3. Generally an increase of the hardness with the number of ARB cycles can be observed. The largest increase is found for the sheet plane, which is due to the high friction between the sheets and the rolls. This leads to a large shear strain at the surface regions of the sheet [17] and, therefore, to a 15% higher hardness compared to the other planes. Moreover, the hardness of both alloys increases severely during the first cycle of ARB, but only slightly during subsequent cycles. While the hardness of X4CrNi18-12 increases rather constantly between one and three cycles, the hardness of X8CrMnNi19-6-3 already appears to saturate after two ARB cycles.

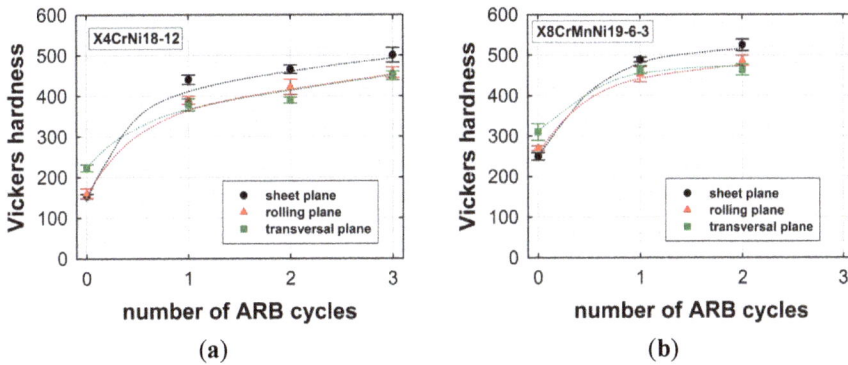

Figure 1. Vickers hardness measurements at the sheet, rolling and transversal planes of the samples after different cycles of ARB for (**a**) X4CrNi18-12 and (**b**) X8CrMnNi19-6-3.

The mechanical performance of both austenitic steels during tensile testing is shown in Figure 2. The ultimate tensile strength (UTS) and the yield strength (YS) increase significantly during the first ARB cycle and the general trend is similar to the hardness measurements. That is to say, both steels reach a YS of around 1.25 GPa after N2 and N3, respectively. The highest increase in strength is found after the first cycle. Therefore, X4CrNi18-12 shows an increase of the UTS by a factor of 1.9 and an increase of the YS by a factor of 3.8. During subsequent ARB-cycles, the strength is further increased, although the relative increase is smaller. The total increase in yield strength compared to the initial material equals a factor of 4.7. The uniform elongation is reduced from 50.4% to 1.2%, while the elongation to failure is reduced from 58% to 7.7%. Similar behavior is found for X8CrMnNi19-6-3, which also shows a strong increase in strength and a strong decrease in ductility after the first ARB cycle. Thereby, the YS increases by a factor of 2.4. The reduction in ductility can be attributed to a decreased hardening rate after severe plastic deformation. Nevertheless, both materials show an excellent combination of strength and ductility. The ARB-processed X8CrMnNi19-6-3 especially performs very well and reaches an UTS between 1.3 and 1.5 GPa, while it maintains an elongation to failure between 10 and 14%. Generally, the shape of the stress-strain curves changes after the first ARB cycle. While the N0 samples deform mainly uniformly during the tensile testing, the ARB-processed ones start necking at pretty low strains but show long post-necking deformation. This transition of tensile

Metals **2015**, *5*, 730–742

deformation behaviors in ultrafine-grained materials was discussed in detail for aluminum by Yu *et al.* [18]. Referring to this publication, tensile stress-strain curves can be categorized into four different characteristic types, in dependence of the grain size and the testing temperature. The N0-curves obtained in the present study a can be assigned to Type IV, which means that the curve shows continuous strain-hardening. This is typically observed in coarse-grained materials with a grain size larger than 4 μm. The stress-strain curves of the ARB processed X4CrNi18-12 can be assigned to Type II, which means that the curves exhibit a distinct yielding peak followed by strain-softening. Type II behavior is observed for grain sizes between 0.4 μm and 1 μm. According to Yu *et al.* [18], the yielding peak is brought into connection with a lack of mobile dislocations, due to the large dislocation sink area provided by grain boundaries. This so-called yield-drop was also found for ultrafine-grained aluminum AA1100 and IF-steel [19], UFG Cu [20], UFG Ti [21] and also for cold-rolled high-manganese austenitic steel [22], which shows twinning induced plasticity. Concerning X8CrMnNi19-6-3, also a distinct yield point followed by strain softening can be observed. In Figure 2d, the true strain hardening rate is determined between the uniform elongation and the elongation to failure and plotted over the true plastic strain for both austenitic steels after N2. The curves were obtained by calculating the true stress strain curves (Figure 2c) from the engineering data and by determining the derivative of those curves. It has to be considered, that the actual cross-section of the tensile samples during tensile testing was not measured. However, the strain hardening curves can be compared qualitatively. It can be observed, that both materials initially show the same behavior up to 5% of plastic strain. That is to say, the strain hardening rate is reduced between the yield point and 3% of true plastic strain. Up to 5% the hardening rate for both alloys is increasing again. Afterwards, the strain-hardening rate of X4CrNi18-12 drastically decreases, while for X8CrMnNi19-6-3 the strain hardening rate increases until it suddenly drops as soon as the sample breaks. This is most likely due to failure because of ARB-related bonding defects. On the one hand, the stabilization of the stress-strain curves could be influenced by an enhanced strain rate sensitivity, which is typically observed for ultrafine-grained fcc metals. Thereby, thermally activated annihilation of dislocations is assumed to play a decisive roll and might lead to increased post-necking strains. On the other hand, both steels show a three-stage work hardening behavior, which is typical for materials with pronounced twinning activity and which is also found for high manganese TWIP (twinning induced plasticity) steels. It appears that the second stage is more distinct for X8CrMnNi19-6-3 compared to X4CrNi18-12. This might be due to the elevated manganese content, which leads to a higher twinning activity during tensile deformation that could stabilize the deformation behavior. The decreasing hardening rate after 5% of plastic strain, which is observed for X4CrNi18-12, might be attributed to a saturation in the amount of twinned grains. This saturation is not reached for X8CrMnNi19-6-3 during plastic straining. However, also martensitic transformation might contribute to the stabilization of the stress-strain curves. To gain more insight, the strain rate sensitivity and the twinning behavior of the materials were investigated.

The strain rate sensitivity (SRS) of the austenitic steels was determined from tensile testing experiments at strain rates of $10^{-3}s^{-1}$, $10^{-4}s^{-1}$ and $10^{-5}s^{-1}$. SRS has to be determined under constant microstructural conditions. Thus, the determination of the SRS from tensile tests is rather crucial, as a microstructural stable condition is hardly achieved. In order to minimize this problem, all stress-strain curves were analyzed at maximum stress, which appears to be a good compromise between the evolution of the microstructure and the limitations of the onset of necking. Therefore, true stress–true strain diagrams were plotted and the maximum stress was evaluated from the different curves. Afterwards, the determined values where plotted over the corresponding strain rate according to [23,24]. Figure 3 reveals that in the initial N0 condition X4CrNi18-12 shows a small positive SRS of around 0.007 at room temperature, while the SRS of X8CrMnNi19-6-3 is around zero. After the first and the second ARB cycle, the SRS of X4CrNi18-12 is slightly decreased to around zero. However, it increases distinctly during the third ARB cycle to 0.017. A similar trend can be found for X8CrMnNi19-6-3, where the SRS remains around zero after the first cycle, but increases after the second cycle to 0.021. The SRS after two cycles in the case of X8CrMnNi19-6-3 and after three cycles in the case of X4CrNi18-12 are in the range of SRS found for other ultrafine-grained fcc materials in literature [6,9,25]. Pronounced strain rate sensitivity is frequently brought into connection with an increased fraction of high angle boundaries, which is typically found in ultrafine-grained fcc metals, see for example [4]. Those high angle grain boundaries can act as sources and sinks for dislocations [3], which are able to stabilize the stress-strain behavior and eventually lead to a higher elongation to failure [6,25]. Consequently, Figure 2 reveals a higher elongation to failure for X8CrMnNi19-6-3 compared to X4CrNi18-12, as the strain rate sensitivity was determined to be higher. However, the SRS usually increases with the number of ARB cycles. Nevertheless, the SRS of X4CrNi18-12 is decreased after the first ARB cycle, which is a rather untypical behavior. In order to clarify more about this point, the microstructures of both alloys were investigated in detail by means of SEM and TEM.

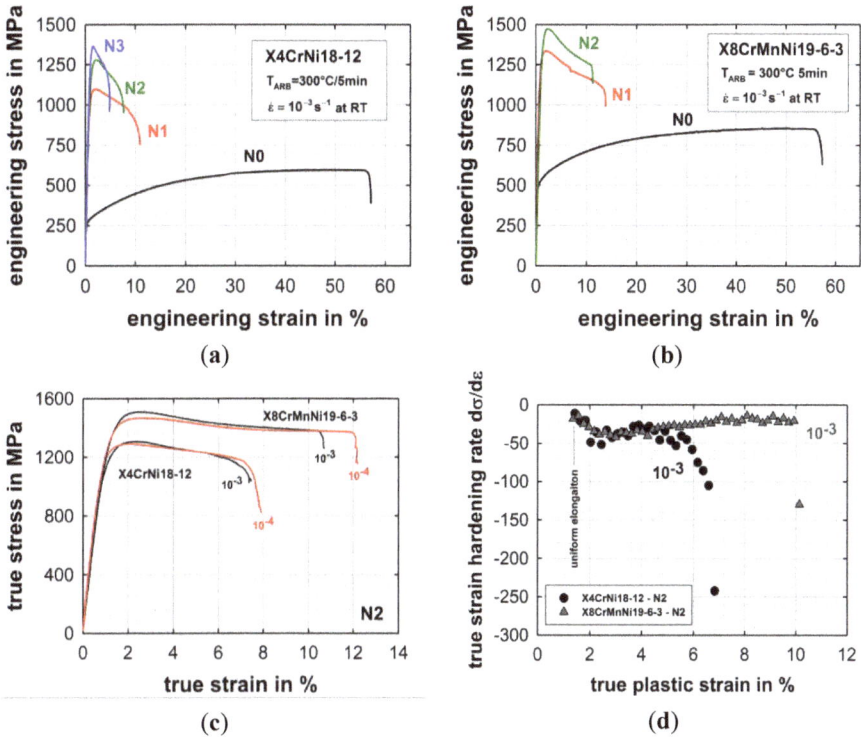

Figure 2. Engineering stress-strain curves at RT and a strain rate of $10^{-3}s^{-1}$ for (**a**) X4CrNi18-12 and (**b**) X8CrMnNi19-6-3. (**c**) Comparative plot of true stress-strain curves of X4CrNi18-12 and X8CrMnNi19-6-3 at strain rates of $10^{-3}s^{-1}$ and $10^{-4}s^{-1}$. Stress-strain curves for $10^{-5}s^{-1}$ are omitted for reasons of clarity. (**d**) True strain hardening rate for X4CrNi18-12 and X8CrMnNi19-6-3 during tensile testing at a strain rate of $10^{-3}s^{-1}$ after N2 cycles of ARB between the uniform elongation and the elongation to failure.

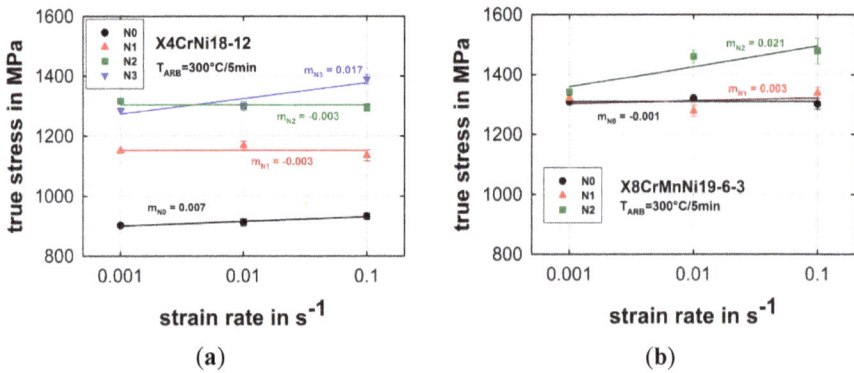

Figure 3. True maximum stress in MPa *vs.* strain rate for (**a**) X4CrNi18-12 and (**b**) X8CrMnNi19-6-3. The slope of the curves equals the strain rate sensitivity m_{Nx}.

Concerning the stabilized austenitic steel X4CrNi18-12 in the initial N0 condition, which can be seen in Figure 4a, the grain size of the equiaxial austenite grains is found to be in the range between 15–30 µm. Moreover, various annealing twins can be observed. The coarse grain structure leads to a high strain hardening capability and therefore to a high uniform elongation. Figure 4b reveals that after one ARB-cycle ($\varepsilon = 0.8$) the grains appear highly deformed and pronounced mechanical twinning and martensite formation are clearly visible. Generally, the microstructure shows various types of microstructural features. On the one hand, large grains with a grain size clearly above 1 µm. These grains are filled up by nano-twins and martensite needles. On the other hand, shear band like regions were found, indicating highly localized plastic deformation with very small grain sizes in the UFG regime. With increasing number of ARB cycles, the fraction of the UFG area is extended and the grain size is further reduced far below 1 µm, see Figure 4c. Moreover, the heterogeneity of the microstructure related to the above described features decreases. Nevertheless, some of the ultrafine grains still contain nano-sized twins in the range of 5–20 nm, which could be observed in the TEM, see Figure 4d. According to Mangonon *et al.* [26], martensitic transformations in Fe-Cr-Ni alloys proceed in the sequence γ (fcc)→ε (hcp)→α' (tetragonal-bcc). Thereby, α' is preferably built at intersections of two ε bands and near regions where ε bands are adjacent to twins or grain boundaries. Furthermore, α' first appears to be needle-shaped and becomes more lath-shaped during subsequent growth. Hereby, the growth of α' leads to a reduction of the ε phase. Shen *et al.* [27] found that both ε and twins act as an intermediate phase during the transformation form γ-austenite to α'-martensite. Above a certain strain level, they observed that the twin density is decreasing, while the martensite density is further increasing. Huang *et al.* [28] performed detailed TEM-investigations on the nucleation mechanism of deformation-induced martensite in an austenitic steel under ECAP-deformation. They observed similar behavior as described above; however they could also find α' nucleating at the intersection of two deformation twins and micro shear-bands. Accordingly, the diffraction patterns of those zones become rather complex (compare inset of Figure 5d).

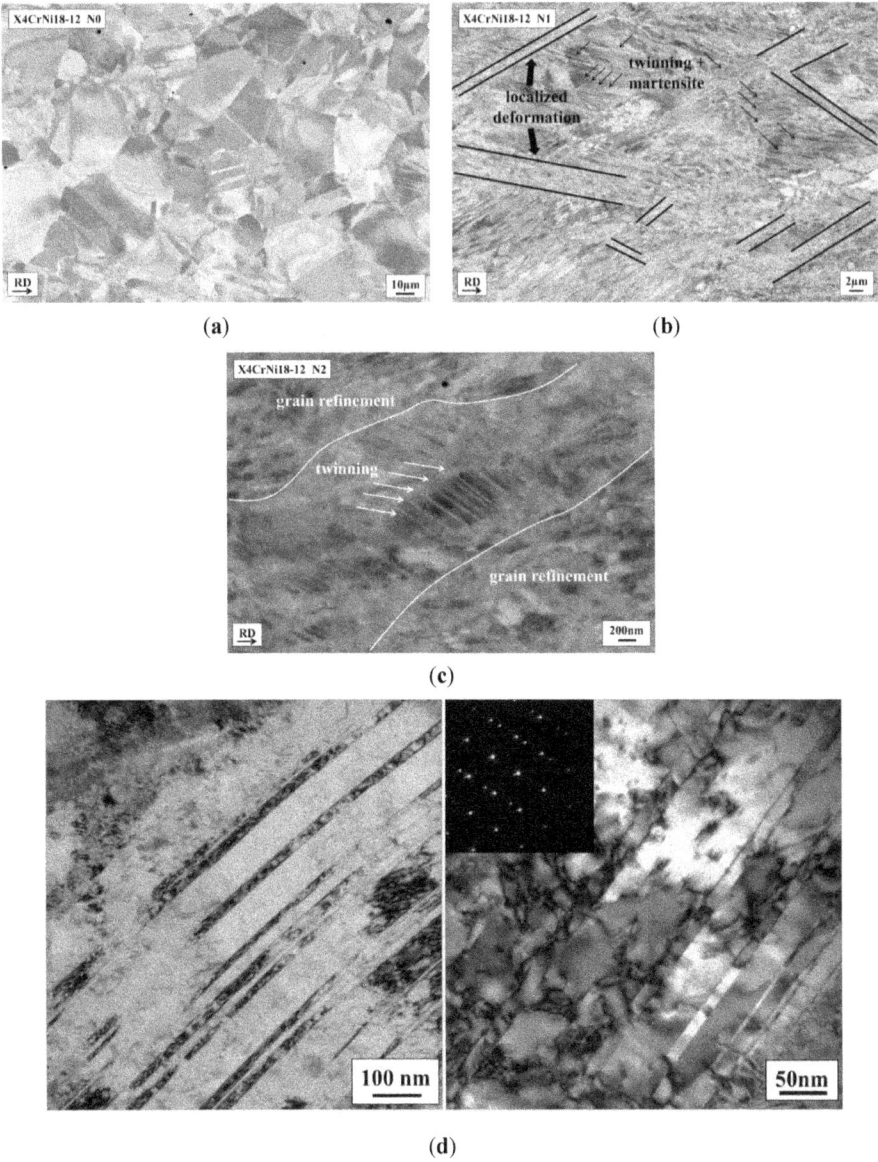

Figure 4. SEM micrographs of ARB-processed austenitic steel X4CrNi18-12 after (**a**) N0, (**b**) N1 and (**c**) N2. The arrows indicate areas where twins but also martensite were found. In (**d**) TEM captions of a highly twinned region with the corresponding diffraction pattern is shown after N2 cycles of ARB.

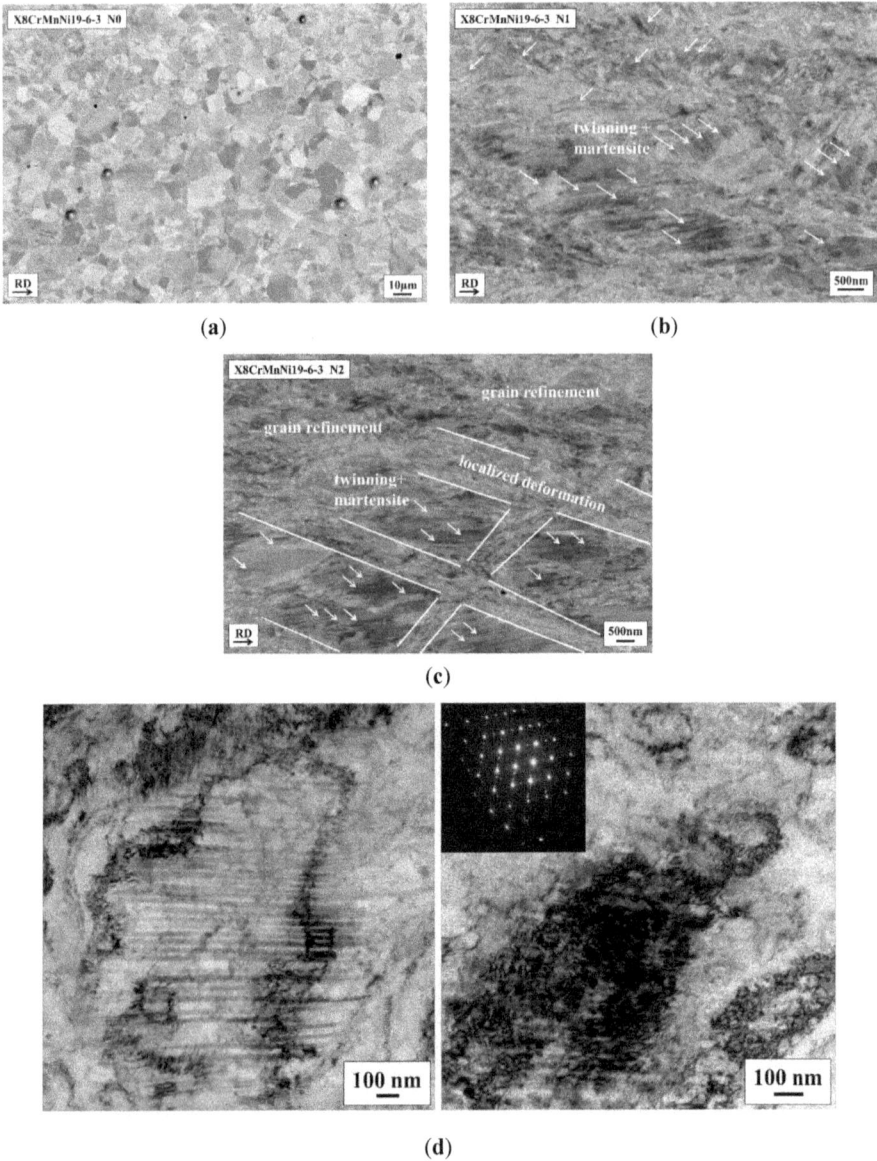

(a)

(b)

(c)

(d)

Figure 5. SEM micrographs of ARB-processed austenitic steel X8CrMnNi19-6-3 after (a) N0, (b) N1 and (c) N2 cycles. The arrows indicate areas where twins but also martensite was found. (d) TEM bright-field caption after N2 ARB-cycles and the corresponding diffraction pattern of this area.

Generally, the microstructural evolution of the austenitic steel X8CrMnNi19-6-3 is quite similar to that of X4CrNi18-12. That is to say, X8CrMnNi19-6-3 also shows equiaxed coarse grains in the initial N0 condition. The grain size is about 10 µm, which is slightly smaller than that of X4CrNi18-12 and there is also a smaller amount of annealing twins, see Figure 5a. After the first ARB cycle, the microstructure already appears highly deformed, and both twinning and martensite are found in the whole microstructure, which can be seen in Figure 5b. Furthermore, some larger grains in the range between 0.5 and 1.5 µm are found, which are rather randomly surrounded by very small grains. After the second ARB cycle, oval-shaped areas with a diameter of 2–5 µm are forming, see Figure 5c. Those areas are divided by zones of localized plastic deformation and very small grain sizes, which appear to be shear bands orientated 45° to rolling direction. Moreover, the fraction of ultrafine-grained areas is clearly increased compared to the N1 condition. In TEM investigations γ-austenite, martensitic areas as well as twinning were found. Accordingly, in [28], the observed diffraction patterns were quite complex. In Figure 5d the diffraction pattern of such an area is representatively shown. Although indications for twinning and martensite could be found, the heavily deformed structures did not allow clear identification and separation of the different phases.

After the first ARB cycle, the microstructure of both alloys is dominated by larger grains containing mechanical twins and martensitic needles. As the fraction of ultrafine grains is rather small, twinning and martensitic transformation can be regarded as the dominating deformation mechanisms. Thus, the highly deformed microstructure contributes to the pronounced increase in strength. Twinning is also assumed to significantly contribute to plastic deformation during tensile testing, which stabilizes the stress-strain behavior. As a consequence of that, elongation to failure becomes relatively high, see Figure 2. Because of pronounced twinning after one ARB cycle and reduced interaction of dislocations with grain boundaries the strain rate sensitivity of both austenitic steels is approaching zero after the first ARB cycle, as shown in Figure 3. This appears to be in contrast to observations by Lu *et al.* [29], who report a positive effect of twins on the SRS in UFG Cu. Nevertheless, both the grain size and the twin thickness are in a much smaller regime in that study. Moreover. martensite formation found in this work might affect the SRS. As the fraction of ultrafine-grained regions is increasing again during subsequent ARB cycles and the twinned areas are reducing, the dislocation grain boundary interaction is enhanced and the strain rate sensitivity is also increasing again. This leads to a positive strain rate sensitivity, which is usually observed in many fcc metals. Mechanisms based on the thermally activated interaction of dislocations with grain boundaries, which might provide a good explanation for the observed increase in SRS, are given by Blum *et al.* [3] in terms of thermally activated annihilation of dislocations at grain boundaries or by Kato *et al.* [30] in terms of thermally activated dislocation depinning at grain boundaries. Moreover, grain boundary sliding cannot be completely neglected and might also contribute to the enhanced SRS to a certain extent. However, when ductility is regarded, the situation becomes more complex: On the one hand, with increasing number of ARB cycles, decreasing elongation to failure is observed as the twinning capability is further decreased and the fraction of martensitic areas becomes more pronounced. On the other hand, the SRS, which is known to positively affect ductility, see Figure 2c, is increased for both alloys. Consequently, the elongation to failure is still rather satisfactory. Therefore, it is assumed that the increased strain rate sensitivity, martensitic transformation as well as twinning appear to contribute to the good ductility during tensile testing.

4. Conclusions

In the present study, commercially available austenitic steels X4CrNi18-12 and X8CrMnNi19-6-3 were processed by accumulative roll bonding at elevated temperatures. For both alloys, pronounced twinning and needles like martensite were found after the first ARB cycle. An increase in yield strength by a factor of 3.9 and 2.4, respectively, was found. During subsequent ARB cycles, the fraction of ultrafine-grained microstructure was clearly increased and a very high yield strength of 1.25 GPa was achieved. For X8CrMnNi19-6-3, the high strength was combined with a satisfactory ductility of more than 10%. The ultrafine-grained regions appeared in channels, dividing coarser areas where pronounced twinning with twin size of 5–20 nm was found. Typical for ultrafine-grained fcc metals, positive strain rate sensitivity between 0.017 and 0.021 was revealed for both alloys, as soon as the fraction of ultrafine-grained regimes was increased. Strain rate sensitivity, twinning and martensitic transformation appear to contribute to the satisfactory ductility.

Metals **2015**, *5*, 730–742

Acknowledgments: The authors would like to thank the German Research Council (DFG) for their financial support and the Cluster of Excellence "Engineering of Advanced Materials", which is funded within the framework of its "Excellence Initiative".

Author Contributions: Mathis Ruppert and Thomas Wenzl performed the mechanical characterization and the SEM investigations. Lisa Patricia Freund prepared the TEM images. Heinz Werner Höppel and Mathias Göken supervised the work and discussed the results with the other authors.

Conflicts of Interest: The authors declare no conflict of interest.

References

1. Saito, Y.; Tsuji, N.; Utsunomiya, H.; Sakai, T.; Hong, R.G. Ultra-fine grained bulk aluminum produced by accumulative roll-bonding (ARB) process. *Scr. Mater.* **1998**, *39*, 1221–1227. [CrossRef]
2. Huang, X.; Tsuji, N.; Hansen, N.; Minamino, Y. Microstructural evolution during accumulative roll-bonding of commercial pure aluminum. *Mater. Sci. Eng. A* **2003**, *340*, 265–271. [CrossRef]
3. Blum, W.; Zeng, X.H. A simple dislocation model of deformation resistance of ultrafine-grained materials explaining Hall-Petch strengthening and enhanced strain rate sensitivity. *Acta Mater.* **2009**, *57*, 1966–1974. [CrossRef]
4. Hughes, D.A.; Hansen, N. High angle boundaries formed by grain subdivision mechanisms. *Acta Mater.* **1997**, *45*, 3871–3886. [CrossRef]
5. Saito, Y.; Utsunomiya, H.; Tsuji, N.; Sakai, T. Novel ultra-high straining process for bulk materials—Development of the accumulative roll bonding (ARB) process. *Acta Mater.* **1999**, *47*, 579–583. [CrossRef]
6. Höppel, H.W.; May, J.; Göken, M. Enhanced strength and ductility in ultrafine-grained aluminium produced by accumulative roll bonding. *Adv. Eng. Mater.* **2004**, *6*, 781–784. [CrossRef]
7. Valiev, R.Z.; Alexandrov, I.V.; Zhu, Y.T.; Lowe, T.C. Paradox of strength and ductility in metals processed by severe plastic deformation. *J. Mater. Res.* **2002**, *17*, 5–8. [CrossRef]
8. Wei, Q. Strain rate effects in the ultrafine grain and nanocrystalline regimes-influence on some responses. *J. Mater. Sci.* **2007**, *42*, 1709–1727. [CrossRef]
9. May, J.; Höppel, H.W.; Göken, M. Strain rate sensitivity of ultrafine-grained aluminium processed by severe plastic deformation. *Scr. Mater.* **2005**, *53*, 189–194. [CrossRef]
10. Tsuji, N.; Ueji, Y.; Minamino, Y. Nanoscale crystallographic analysis of ultrafine grained IF steel fabricated by ARB process. *Scr. Mater.* **2002**, *47*, 69–76. [CrossRef]
11. Kamikawa, N.; Sakai, T.; Tsuji, N. Effect of redundant shear strain on microstructure and texture evolution during accumulative roll-bonding in ultralow carbon IF steel. *Acta Mater.* **2007**, *55*, 5873–5888. [CrossRef]
12. Lee, S.-H.; Saito, Y.; Park, K.-T.; Shin, D.H. Microstructures and mechanical properties of ultra low carbon IF steel processed by accumulative roll bonding process. *Mater. Trans.* **2002**, *43*, 2320–2325. [CrossRef]
13. Kitahara, H.; Tsuji, N.; Minamino, Y. Martensite transformation from ultrafine grained austenite in Fe-28.5 at. % Ni. *Mater. Sci. Eng. A* **2006**, *438–440*, 233–236. [CrossRef]
14. Jafarian, H.; Eivani, A. Texture development and microstructure eolution in metastable austenitic steel processed by accumulative roll bonding and subsequent annealing. *J. Mater. Sci.* **2014**, *49*, 6570–6578. [CrossRef]
15. Shen, Y.F.; Zhao, X.M.; Sun, X.; Wang, Y.D.; Zuo, L. Ultrahigh strength of ultrafine grained austenitic stainless steel induced by accumulative rolling and annealing. *Scr. Mater.* **2014**. [CrossRef]
16. Li, B.; Tsuji, N.; Minamino, Y. Microstructural evolution in 36%Ni austenitic steel during the ARB process. *Mater. Sci. Forum* **2006**, *512*, 73–78. [CrossRef]
17. Lee, S.-H.; Saito, Y.; Tsuji, N.; Utsunomiya, H.; Sakai, T. Role of shear strain in ultragrain refinement by accumulative roll-bonding (ARB) process. *Scr. Mater.* **2002**, *46*, 281–285. [CrossRef]
18. Yu, C.Y.; Kao, P.W.; Chang, C.P. Transition of tensile deformation behaviors in ultrafine-grained aluminum. *Acta Mater.* **2005**, *53*, 4019–4028. [CrossRef]
19. Tsuji, N.; Ito, Y.; Saito, Y.; Minamino, Y. Strength and ductility of ultrafine grained aluminum and iron produced by ARB and annealing. *Scr. Mater.* **2002**, *47*, 893–899. [CrossRef]

20. An, X.H.; Wu, S.D.; Zhang, Z.F.; Figueiredo, R.B.; Gao, N.; Langdon, T.G. Enhanced strength-ductility synergy in nanostructured Cu and Cu-Al alloys processed by high-pressure torsion and subsequent annealing. *Scr. Mater.* **2012**, *66*, 227–230. [CrossRef]

21. Li, Z.; Fu, L.; Fu, B.; Shan, A. Yield point elongation in fine-grained titanium. *Mater. Lett.* **2013**, *96*, 1–4. [CrossRef]

22. Saha, R.; Ueji, R.; Tsuji, N. Fully recrystallized nanostructure fabricated without severe plastic deformation in high-Mn austenitic steel. *Scr. Mater.* **2013**, *68*, 813–816. [CrossRef]

23. Hart, E.W. Theory of the tensile test. *Acta Metall.* **1967**, *15*, 351–355. [CrossRef]

24. Ghosh, A.K. On the measurement of strain-rate sensitivity for deformation mechanism in conventional and ultra-fine grain alloys. *Mater. Sci. Eng. A* **2007**, *463*, 36–40. [CrossRef]

25. Höppel, H.W.; May, J.; Eisenlohr, P.; Göken, M. Strain-rate sensitivity of ultrafine-grained materials. *Zeitschrift für Metallkunde* **2005**, *96*, 566–571. [CrossRef]

26. Mangonon, P.; Thomas, G. The martensite phases in 304 stainless steel. *Metall. Trans.* **1970**, *1*, 1577–1586. [CrossRef]

27. Shen, Y.F.; Li, X.X.; Sun, X.; Wang, Y.D.; Zuo, L. Twinning and martensite in a 304 austenitic stainless steel. *Mater. Sci. Eng. A* **2012**, *552*, 514–522. [CrossRef]

28. Huang, C.X.; Yang, G.; Gao, Y.L.; Wu, S.D.; Li, S.X. Investigation on the nucleation mechanism of deformation-induced martensite in an austenitic stainless steel under severe plastic deformation. *J. Mater. Res.* **2007**, *22*, 724–729. [CrossRef]

29. Lu, K.; Lu, L.; Suresh, S. Strengthening materials by engineering coherent internal boundaries at the nanoscale. *Science* **2009**, *324*, 349–352. [CrossRef] [PubMed]

30. Kato, M. Thermally activated dislocation depinning at a grain boundary in nanocrystalline and ultrafine-grained materials. *Mater. Sci. Eng. A* **2009**, *516*, 276–282. [CrossRef]

metals

MDPI

Article

Fatigue Behavior of Ultrafine-Grained Medium Carbon Steel with Different Carbide Morphologies Processed by High Pressure Torsion

Christoph Ruffing [1], Aaron Kobler [2,3], Eglantine Courtois-Manara [2], Robby Prang [2], Christian Kübel [2,4], Yulia Ivanisenko [2] and Eberhard Kerscher [1,*

[1] Materials Testing, University Kaiserslautern, Gottlieb-Daimler-Straße, 67663 Kaiserslautern, Germany; ruffing@mv.uni-kl.de

[2] Institute of Nanotechnology (INT), Karlsruhe Institute of Technology (KIT), Hermann-von-Helmholtz-Platz 1, 76344 Eggenstein-Leopoldshafen, Germany; aaron.kobler@kit.edu (A.K.); eglantine.courtois-manara@kit.edu (E.C.-M.); robby.prang@t-online.de (R.P.); christian.kuebel@kit.edu (C.K.); julia.ivanisenko@kit.edu (Y.I.)

[3] Joint Research Laboratory Nanomaterials (KIT and TUD), Technische Universität Darmstadt (TUD), Petersenstr. 32, 64287 Darmstadt, Germany

[4] Karlsruhe Nano Micro Facility (KNMF), Karlsruhe Institute of Technology (KIT), Hermann-von-Helmholtz-Platz 1, 76344 Eggenstein-Leopoldshafen, Germany

* Author to whom correspondence should be addressed; kerscher@mv.uni-kl.de; Tel.: +49-631-205-2136; Fax: +49-631-205-5261.

Academic Editor: Heinz Werner Höppel
Received: 21 March 2015; Accepted: 25 May 2015; Published: 29 May 2015

Abstract: The increased attention ultrafine grained (UFG) materials have received over the last decade has been inspired by their high strength in combination with a remarkable ductility, which is a promising combination for good fatigue properties. In this paper, we focus on the effect of different carbide morphologies in the initial microstructure on the fatigue behavior after high pressure torsion (HPT) treatment of SAE 1045 steels. The two initial carbide morphologies are spheroidized as well as tempered states. The HPT processing increased the hardness of the spheroidized and tempered states from 169 HV and 388 HV to a maximum of 511 HV and 758 HV, respectively. The endurance limit increased linearly with hardness up to about 500 HV independent of the carbide morphology. The fracture surfaces revealed mostly flat fatigue fracture surfaces with crack initiation at the surface or, more often, at non-metallic inclusions. Morphology and crack initiation mechanisms were changed by the severe plastic deformation. The residual fracture surface of specimens with spheroidal initial microstructures showed well-defined dimple structures also after HPT at high fatigue limits and high hardness values. In contrast, the specimens with a tempered initial microstructure showed rather brittle and rough residual fracture surfaces after HPT.

Keywords: severe plastic deformation; high pressure torsion; fatigue; carbide morphology; shear bands; high strength steels; microstructure; fracture surface

1. Introduction

Despite the progress achieved over the last fifty years in the development of new steel grades and thermal- and thermomechanical treatments, the quest for novel processing routes allowing further enhancement of mechanical properties remains of great current interest. Furthermore, the field of nanostructuring has not been extensively explored in mainstream steel research. It was proposed that grain size refinement could be the most promising way to improve the fatigue life of steel because it allows obtaining high strength in combination with good ductility values [1,2]. Severe plastic deformation (SPD) of metals and alloys is a well-established method to obtain ultrafine-grained structures, or phase compositions that are impossible to obtain by conventional thermal treatment. Essential for SPD is the combination of a high hydrostatic pressure, to avoid crack initiation, and an enormous shear strain. At present, the most developed SPD techniques are Equal Channel Angular Pressing (ECAP) [3], High Pressure Torsion (HPT) [4] and accumulated Roll Bonding (ARB) [5]. The possibility of producing bulk nanostructured metals with grain sizes in the nanometre to submicrometre ranges has been demonstrated using these methods [6]. In the case of multiphase alloys and intermetallic compounds, SPD not only leads to grain refinement but can also lead to the formation of non-equilibrium solid solutions [2,7], disordering [8], or amorphization [9].

Recent investigations of the mechanical behaviour of ultrafine-grained materials (nano- and submicrocrystalline) processed by severe plastic deformation have demonstrated a significant improvement of their strength compared to their coarse-grained counterparts [6,10,11]. Furthermore, high cycle fatigue properties were substantially improved in several face centered cubic (fcc) metals with grain reduction down to UFG and nanoscopic scales [12,13]. Similar results were also obtained for ferritic and low carbon steels [14–16]. In particular, in low carbon steel C15 processed by ECAP, the increase of the endurance limit by a factor of two has been reported [17]. Nevertheless, the fatigue properties characteristic for martensitic or bainitic steels have not been achieved yet. One of the major problems associated with UFG structures produced by SPD is their inherent thermal, structural, and mechanical instability, caused by the high dislocation density, local internal stresses created during cold working, and limited hardenability. For example, a large amount of low angle boundaries introduced by ECAP processing led to cyclic instability in low carbon steel C10 [18].

In this paper, we present fatigue properties of ultrafine-grained medium carbon steels with two different carbide morphologies. In order to obtain microstructures with a high fraction of high angles grain boundaries and low levels of internal stress, we applied HPT processing at elevated temperatures–warm HPT [19]. First, the paper offers an intensive characterization of the resulting microstructure after HPT treatment. Automated crystal orientation mapping in a transmission electron microscope (ACOM-TEM) was used in addition to standard scanning electron microscopy (SEM). In the fatigue tests, use was made of electro-dynamic test equipment for micro-bending. We can show a correlation of the hardness, which is affected by the grain refinement during HPT, with the endurance limits, which were determined in the fatigue tests. Fractographic investigations helped to interpret and analyze the results.

2. Experimental Section

The investigations were conducted with commercial medium carbon steel SAE 1045 (Fe-balance, 0.46 C, 0.64 Mn, 0.17 Si, 0.011 P, 0.009 S, all in wt. %) delivered as rods with a normalized ferritic-pearlitic microstructure. The rods were heat treated to obtain two different carbide morphologies: In order to obtain spheroidized carbides, part of the rods was annealed at 680 °C for 40 h (initial state S). The remaining rods were annealed at 850 °C, quenched to room temperature and tempered for 1 h at 450 °C (initial state T). Subsequently, the thermally treated rods were cut into 0.8 mm thin discs and processed by high pressure torsion under a pressure of 6 GPa for six and ten rotations at a temperature of 380 °C. The shear strain γ experienced by HPT processed specimens can be calculated using Equation (1):

$$\gamma = \frac{2\pi n}{t}r \tag{1}$$

with r, n, and t as the distance from the axis of rotation, the number of HPT revolutions, and the thickness of the disc, respectively. The shear strain is 141 after six and 236 after ten rotations at a distance of 3 mm from the center of the disk. This 3 mm distance is also the location of the specimen extraction for structural characterization and hardness measurements.

Four rectangular fatigue samples with dimensions of 4 mm × 1 mm (w) × 0.6 mm (h) were cut with their center at a distance of 3 mm from the center of every 10 mm × 0.6 mm cylindrical HPT disk as illustrated in Figure 1a. The fatigue samples were ground into the final shape and polished to 1 μm grid size. Hardness measurements were carried out with an ASMEC UNAT 2 Nanoindenter (ASMEC GmbH, Radeberg, Germany) and a proof force of 200 mN on a surface polished with colloidal SiO_2.

Cyclic four-point-bending tests were performed on a BOSE ELECTROFORCE 3230 electrodynamic testing machine at a frequency of 40 Hz under stress control. They were conducted with a load ratio of $R = 0.1$ at room temperature. The mountings are shown in Figure 1b with a lower mounting distance of 3.2 mm and an upper one of 1.6 mm. To determine the endurance limits, the staircase method [20] was used with different step sizes, depending on the fatigue limit to be expected by a linear correlation with the hardness. Stress calculations were carried out for the controlled force F considering the specimen geometry, using Equation (2):

$$\sigma_{edge} = \frac{3Fd}{wh^2} \tag{2}$$

In this case, $d = 0.8$ mm is the distance between the upper and lower mounting rollers. The parameter σ_{edge} specifies the edge stress as real stress in longitudinal direction in the lowest tension fiber when linear-elastic material behavior is assumed.

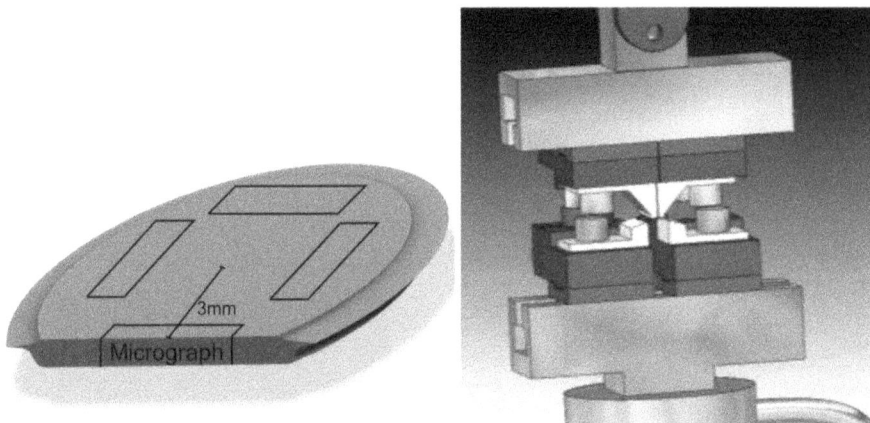

Figure 1. (**a**) HPT disc with indicated specimen extraction areas. (**b**) cyclic four-point-bending test fixture.

A PHILIPS XL40 scanning electron microscope (SEM, Philips, Eindhoven, The Netherlands) was used for fractographic investigations and low magnification microstructure investigations at a voltage of 20 kV. For SEM investigations with higher magnification, a ZEISS Supra 40VP (Carl Zeiss Microscopy GmbH, Oberkochen, Germany) was used at a voltage of 5 kV. The microstructure of the HPT processed samples was further investigated by transmission electron microscopy (TEM). TEM lamellas were cut from the HPT deformed discs at the position where the SEM investigations had been carried out. These areas are located approximately 3 mm from the center (Figure 1). The lamellae were prepared by FIB lift-out using a FEI Strata 400S (FEI, Hillsboro, OR, USA) with the final polishing performed with a 5kV Ga+-ion beam. The TEM lamellas of samples after fatigue fractures were taken from the location of the fracture initiation site.

Automated crystal orientation mapping (ACOM-TEM) was conducted using a Tecnai F20 ST (FEI, Hillsboro, OR, USA) equipped with an ASTAR DigiStar (Nanomegas, Brussels, Belgium) system and operated at 200 kV in μp-STEM mode with spot size 7, camera length 150 mm, condenser aperture

20 µm, gun lens 6 and extraction voltage 4.5 kV [21]. Orientation maps have been obtained with an acquisition speed of 100 frames per second. Data processing for area weighted grain size determination consisted of the following steps:

Crystallite/grain recognition in the orientation maps with an extended version of Mtex 3.4.2 [22]. The chosen disorientation between neighbors for the recognition was 5°.

Pixel filtering of the orientation maps was performed with a median filter of the combined Euler angles, weighted by the phase [23].

Crystallites that exhibit a 180° orientation ambiguity to their neighbors were corrected [24].

- Small grains and grains with a bad reliability and index were removed.
- No re-filling of any of the removed pixels was undertaken.

BF/DF-TEM analysis was carried out with the help of an image corrected FEI Titan 80-300 (FEI, Hillsboro, OR, USA), equipped with a Gatan US1000 (Gatan, Pleasanton, CA, USA) slow scan CCD camera operated at 300 kV.

3. Results and Discussion

3.1. Materials and Microstructure

The microstructure of C45 steel after spheroidizing annealing (S state) is shown in Figures 2a and 3a,b.

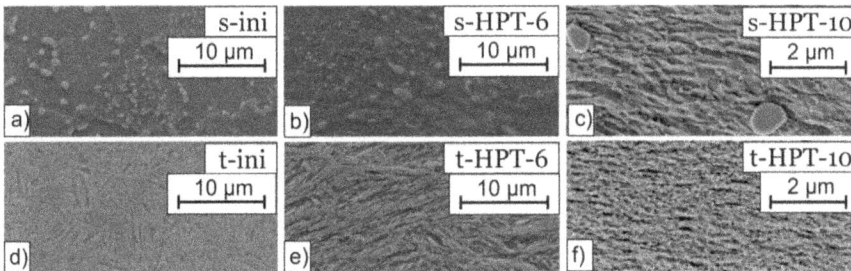

Figure 2. Etched surface microstructures prior to the fatigue tests (**a**) initial state after soft annealing containing spheroidal carbides (**b**) after 6 HPT rotations (**c**) after 10 HPT rotations (**d**) initial state with tempered microstructure (**e**) after 6 rotations (**f**) after 10 rotations.

The soft annealed state contains spheroidal carbides distributed in a uniform coarse grained microstructure with well-defined grain boundaries (Figures 2a and 3b). The grains are micrometer sized (~15 µm) and initial hardness is 169 HV. This state will be referred to as s-ini. Typical images of the S state after warm HPT for six and ten rotations (these states will be referred to as s-HPT-6 and s-HPT-10) are shown in Figure 2b,c and Figure 3c–f. After six rotations, the microstructure is refined and new grain boundaries appear (Figures 2b and 3c). The grain size shrinks to approximately 0.8 µm and the grains show a slight elongation along the HPT shear direction (Figure 3c) with a hardness of 289 HV. After ten rotations, the ferrite grain size decreases further to approximately 0.2 µm and the elongation is more pronounced (Figure 3e,f). Hardness rises to 511 HV. The continuous grain refinement is best visible in Figure 3a,c,e.

The spheroidal carbides are still clearly visible in s-HPT-10, but the larger ones in Figure 3e are partially fragmented and appear blurred. In addition, an increasing number of small carbides are distributed between the ferrite grains.

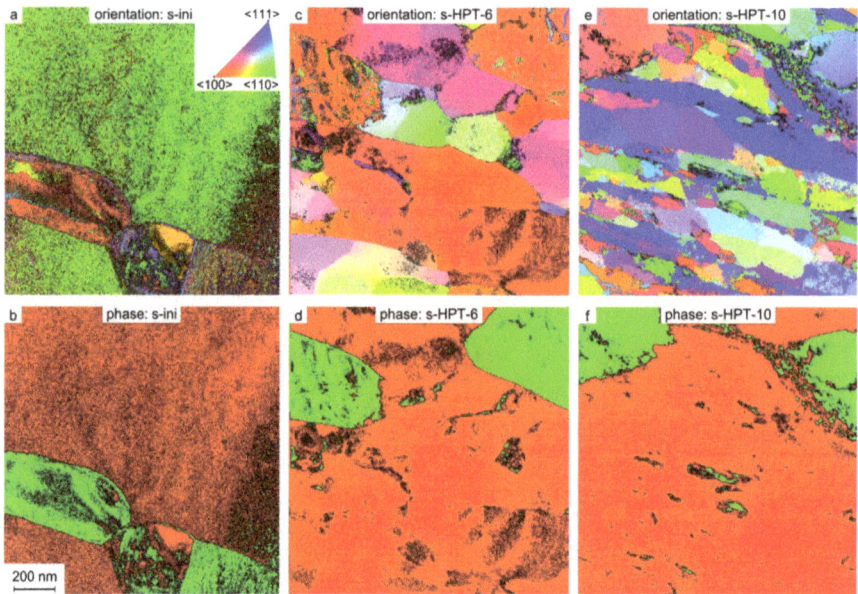

Figure 3. Orientation (**a,c,e**) and phase (**b,d,f**) maps (ACOM-TEM) overlaid with the reliability of the spheroidized microstructures (**a,b**), after 6 (**c,d**) and 10 HPT rotations (**e,f**) all prior to the fatigue tests. The color code of the orientation maps is given in (**a**). The projection direction is normal to the paper plane. The Fe phase is given in red and green represents the carbide phase. The reliability is given in black for both types of maps.

Needle-shaped microstructures are visible in Figures 2d and 4a,b for C45 steel after tempering (T state). Carbides are finely dispersed between the needles (Figure 4b). The samples have an initial hardness of 388 HV before HPT. This state will be referred to as t-ini. After six HPT rotations (t-HPT-6), the needles become slightly refined (Figures 2e and 4c), their long axis is oriented along the shear direction and the formation of new low and high angle grain boundaries is visible. The hardness increases to 457 HV in t-HPT-6. After ten rotations (t-HPT-10), the microstructure shows strong refinement by SEM (Figure 2f) with no visible needles anymore. ACOM-TEM reveals an ultrafine grained structure with elongated grains and the cementite is finely dispersed in the ferrite matrix. The final hardness at the fatigue samples extraction point is 758 HV. However, the microstructure and hardness show variations along the radius of the HPT disk due to the shear strain variation (compare Equation (1)). The cementite distribution is most homogeneous after 10 rotations.

After HPT, shear bands are visible by optical microscopy on the surface, fracture surface, and cross section of all specimens. Such shear bands have been reported earlier for HPT processed nanocrystalline metals such as palladium [25]. Interesting to notice is that, in the present case of the C45 steel, non-metallic inclusion were sometimes sheared by the shear bands as seen in the FIB cross-section in Figure 5.

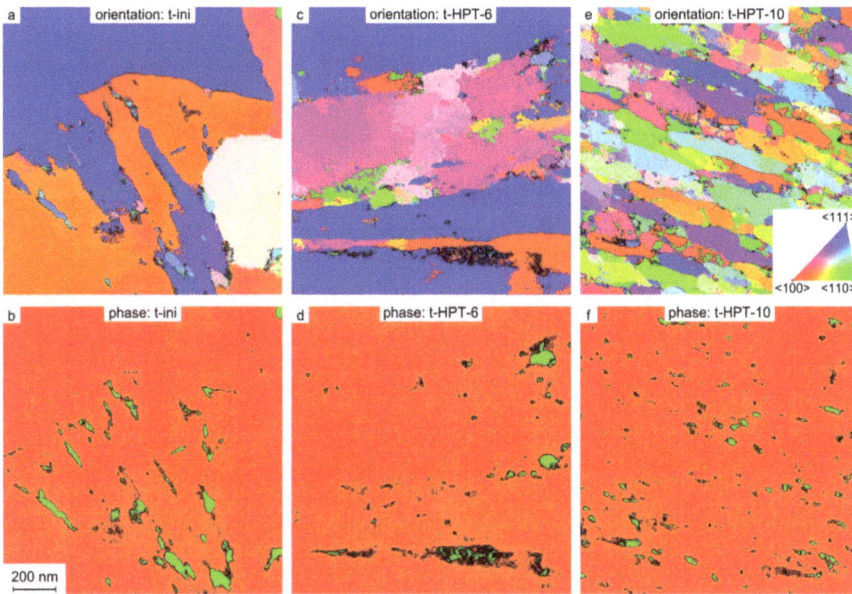

Figure 4. Orientation (**a,c,e**) and phase (**b,d,f**) maps (ACOM-TEM) overlaid with the reliability of the tempered microstructures (**a,b**), after 6 (**c,d**) and 10 HPT rotations (**e,f**) all prior to the fatigue tests. The color code of the orientation maps is given in (**e**). The projection direction is normal to the paper plane. The Fe phase is given in red and green represents the carbide phase. The reliability is given in black for both types of maps.

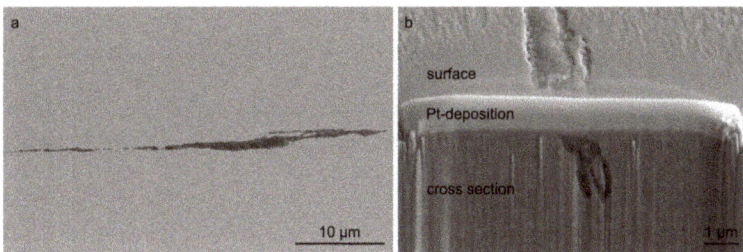

Figure 5. SEM images of a sheared non-metallic inclusion on the (non-etched) surface of t-HPT-10: (**a**) top view; (**b**) FIB cut through the inclusion.

3.2. S-N-Curves

Four-point-bending fatigue tests were carried out with both carbide morphologies in the initial coarse grained (CG) and the HPT deformed states (Figure 6). The edge stress amplitude was plotted against the number of cycles to fracture for both initial carbide morphologies. The arrows in the squares indicate runouts after 10^7 cycles. The small numbers beside the arrows show the number of overlapping runout-points. The crack initiation site type is indicated on top of the diagrams. A few specimens could not be investigated (n.I.) after fatigue, e.g., because they did not break completely.

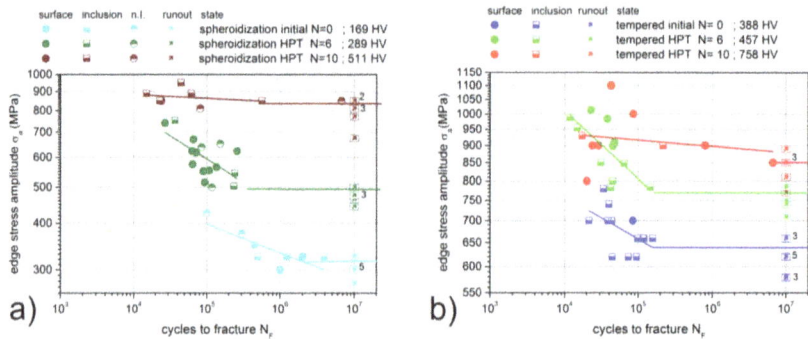

Figure 6. S-N curves of (**a**) CG and UFG spheroidized SAE 1045 revealing endurance limits of 316 MPa, 493 MPa and 837 MPa and of (**b**) UFG and CG tempered SAE 1045 showing endurance limits of 640 MPa, 769 MPa and 850 MPa.

Figure 6a shows the values for the spheroidizing annealed state in the CG (s-ini) and the UFG (s-HPT-6 and s-HPT-10) condition. For the s-ini state, a fatigue limit of 316 MPa was determined by the staircase method. Fractured specimens were observed after more than 10,000,000 cycles at stress amplitude of 325 MPa, while at stress amplitude of 300 MPa only one specimen fractured. In contrast, s-HPT-6 with a hardness of 289 HV showed fatigue fracture at lifetimes below about 400,000 cycles. The endurance limit (494 MPa) was significantly higher compared to s-ini. Crack initiation was mainly at the surface as indicated in the diagram. After 10 HPT rotations, resulting in a hardness of 511 HV, the fatigue limit further increased to 837 MPa. One specimen failed at 800,000 cycles, all other either failed within the first 80,000 cycles or ran out. Below 800 MPa, no fatigue failure was observed for s-HPT-10. The crack initiation site changed with an increasing number of HPT-rotations from the surface without non-metallic inclusions for s-HPT-6 to crack initiation at non-metallic inclusions at the surface in s-HPT-10 (see also Figure 7).

Figure 6b shows the S-N-curves for t-ini, t-HPT-6 and t-HPT-10. A fatigue limit of 640 MPa was determined for t-ini, corresponding to a hardness of 388 HV. In this case, crack initiation mostly occurred at non-metallic inclusions at the surface. With an increased hardness of 457 HV, the fatigue limit of t-HPT-6 rose to 769 MPa. For higher fatigue loads, a specific crack initiation location could not be identified. The exceptions are cyclic edge stresses between 780 MPa and 900 MPa, where all cracks initiated at non-metallic inclusions. For the t-HPT-10 specimens, offering the highest shear strain and hardness, the S-N curves and the crack behavior seem to be completely different. In contrast to the initial state and t-HPT-6, where specimens either failed before 200,000 cycles or ran out, many t-HPT-10 samples failed at a higher number of cycles. In addition, this state also revealed a number of very early cracks, which reduced the fatigue limit to a value of 850 MPa, lower than expected for a hardness of 758 HV, when considering a linear correlation between hardness and fatigue limit. Cracks were mostly initiated at the surface of the t-HPT-10 specimens.

3.3. Fractographic Investigations

All fracture surfaces of the broken specimens were investigated by SEM after the bending fatigue tests. In bending specimens, a stress gradient over the thickness leads to tension on one side, a neutral plane, and compression on the other side. The tension side, where fatigue cracks always start in this kind of material, is always shown at the bottom for the SEM images in this paper.

In the following fractographic investigations, only the HPT-10 states are included because of their large scatter in lifetime compared to all other states.

Figure 7a shows the fracture surface of one of the s-HPT-10 specimens after a fatigue load of 850 MPa. A large fatigue fracture surface (Figure 7b) is visible, which is larger than typically found in conventional high strength materials. The fatigue fracture surface shown in Figure 7b is very homogeneous and appears smooth. The crack initiation occurred at a non-metallic inclusion. In the transition area between fatigue and residual fracture surface, some short cave-like shear bands are visible. Such shear bands are observed in fracture surfaces of HPT processed materials, for example in nanocrystalline palladium after HPT [25]. In this specimen, shear bands seem not to have influenced the crack initiation or early fatigue crack growth because they are observed only in the late fatigue crack growth areas well as in the residual fracture area.

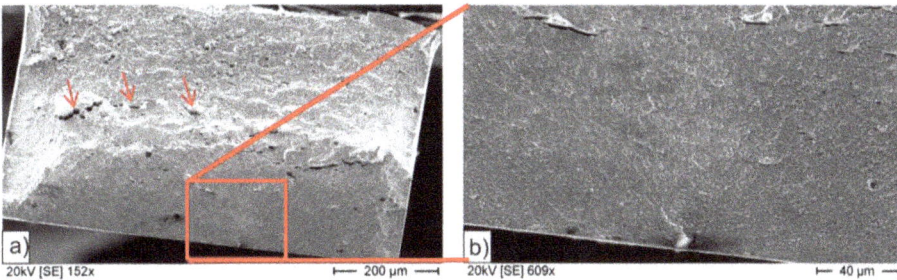

Figure 7. Fracture surface of s-HPT-10 (**a**) whole fracture surface (**b**) fatigue fracture surface; some shear bands resulting from HPT are shown with arrows.

Figure 8 illustrates the fracture surface of a t-HPT-10 sample loaded at a stress of 900 MPa. SEM investigations reveal shear bands in the fracture surface. The cave-like characteristics appear over the whole residual fracture surface. With a length of up to 200 µm, they seem to be larger than those in the spheroidizing annealed state, which are only up to 40 µm long. The fatigue fracture surface in Figure 8b is smaller than in the s-HPT-10 state although the sample was loaded at nearly the same stress level. We also observe crack initiation at a non-metallic inclusion. The fracture appearance is rather homogenous with a flat fatigue fracture surface. In addition, some shear bands are visible in the fatigue fracture surface but similar to the spheroidizing annealed state, they do not influence the crack initiation or propagation.

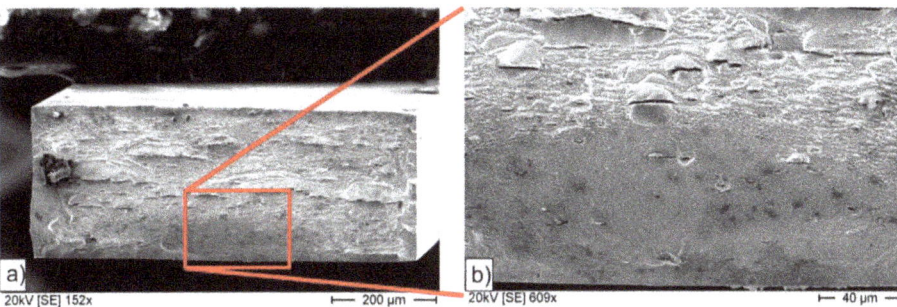

Figure 8. Fracture surface of t-HPT-10 (**a**) whole fracture surface (**b**) crack initiation area at the surface without nonmetallic inclusion and fatigue fracture surface.

Figure 9 shows a comparison of the residual fracture surfaces of the two states with tempered and spheroidizing annealed microstructure after HPT. There are significant differences in the morphology. In Figure 9a, s-HPT-10 shows well-defined dimples, which indicate a homogeneous ductile fracture behavior.

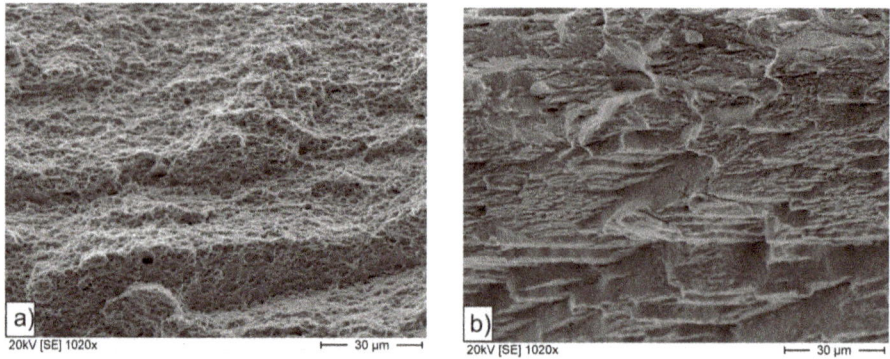

Figure 9. Residual fracture surface of (**a**) s-HPT-10 and (**b**) t-HPT-10.

In contrast, Figure 9b reveals a rather rough residual fracture surface with sharp edges, which are common for this material state and indicate a brittle material behavior. Most likely, these sharp edges are shear steps from shear banding. Figure 10a,b show an FIB cross-section of the t-HPT-residual fracture surface at a characteristic cave, which was formed by a shear band. This indicates a correlation between shear bands observed at the fracture surface and the microstructure underneath. The shear band at the fracture surface is connected to a line-shaped inhomogeneity inside the microstructure. A similar line pattern is also visible in a polished and Nital-etched cross section of the HPT specimen in Figure 10c prior to fatigue loading. These lines are extended along the shear direction during HPT and become visible after etching in SEM in this case in the t-HPT-10 state. However, similar shear bands can also be seen in the micrographs of the s-HPT-10 states.

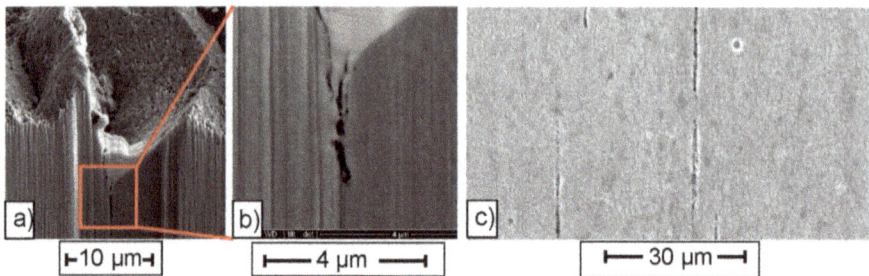

Figure 10. SEM images of shear bands in t-HPT-10 samples (**a**,**b**) inside the residual fracture surface and (**c**) on the polished and etched cross section.

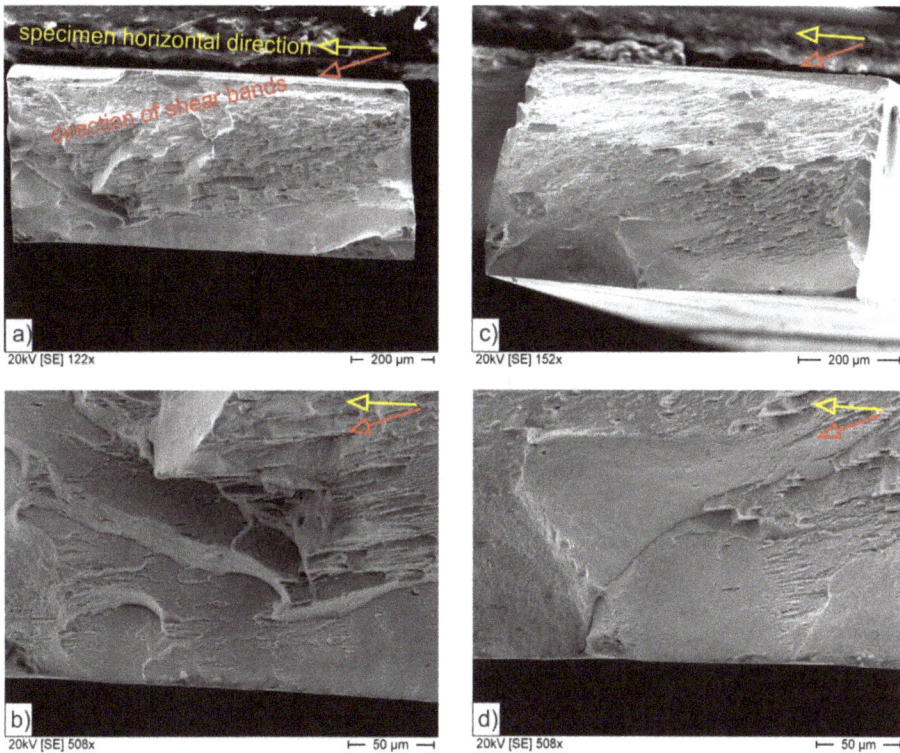

Figure 11. Fracture surface of t-HPT-10 state which shows shear bands (**a,b**) loaded at 900 MPa and (**c,d**) loaded at 850 MPa (red open arrows at the top indicate the direction of shear bands in correlation to the specimens surface indicated by the yellow open arrows. The crack initiation site is always at the bottom of the pictures).

The fracture surface of a tempered specimen after HPT, which failed very early at 900 MPa, is shown in Figure 11a,b. The picture shows a cliff-like fracture surface with a high topology in the residual and also in the fatigue fracture surface. The fatigue crack seems to change direction and jump between different levels. Usually, fatigue cracks grow mostly perpendicular to the tensile stress in one defined direction, as visible in Figures 7 and 8. The fact that this is not the case here is an indication of some inhomogeneities inside the material prior to fatigue loading which affected the crack initiation site and the crack path. The whole fracture surface offers different characteristics of the shear bands compared to the fatigue samples with higher endurance limits. The shear bands seem to influence the crack growth and the lifetime of the fatigue sample because there is a link between the shear bands inside the residual fracture surface and the different levels and layers in the fatigue fracture surface.

A similar behavior is presented in Figure 11c,d, showing the fracture surface of a t-HPT-10 state sample, which cracked after 6,000,000 cycles. It is obvious that shear bands are present in the area of fatigue crack growth, especially at the crack initiation site, where an extension of a shear bands are points of multiple crack initiation. These shear bands have been produced during HPT and seem to have promoted the crack initiation during fatigue load. Further crack growth in the fatigue fracture surface is similar to that shown in Figure 11a,b, exhibiting a high topology. The fatigue crack growth is dominated by the shear bands (marked with red arrows in all pictures). The two arrows in each part of Figure 11 indicate that there is an angle of misorientation between the expected shear plane at HPT (yellow arrows) and the shear band plane in the respective specimen (red arrows). This is the case for both specimens, which both exhibited early failure during fatigue testing, and it establishes a significant difference in contrast to samples in the spheroidization annealed or tempered state, which offer a homogeneous fracture behavior in the S-N curve without early cracks (t-HPT-6). For both specimens shown in Figure 11, a complex structure of different fracture planes is visible in the fatigue fracture surface. The surface of each single plane itself seems to be very flat, as expected for fatigue failure in ultrafine grained materials, resulting in a strong topology with clustering in different spatial directions.

Figure 12 shows the crack initiation site marked in Figure 11c,d in higher magnification after rotating the specimen by 20° out of plane. Some penetration lines are visible at the specimen surface where shear bands from HPT deformation seem to have been re-activated during fatigue loading. The assumed path of the shear bands inside the specimen is indicated by parallel lines, which fits the direction estimated from the overall fracture surface in Figure 11.

20kV [SE] 777x   ⊢ 30 µm ⊣

Figure 12. Shear bands visible at the sample's surface in the area of the crack initiation site.

Figure 13 shows cross sections of typical crack initiation sites for the HPT states after ten HPT rotations. BF-TEM and ACOM-TEM images were obtained to correlate the crack path, which is in both cases horizontal in the upper part of the images, with the microstructure underneath. In Figure 13a, a non-metallic inclusion, located on the left side of the shown area, was responsible for crack initiation in the s-HPT-10 state. Grain size and morphology close to the crack do not differ significantly from the pre-fatigue s-HPT-10 state.

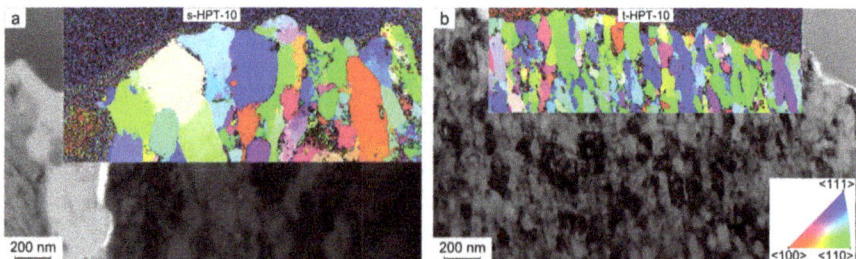

Figure 13. Cross section of the crack initiation (**a**) non-metallic inclusion for s-HPT-10 and (**b**) crack path near crack initiation from the surface for t-HPT-10. The overlays are orientation maps in projection direction normal to the paper plane with the reliability in black. The color code of the orientation maps is given in (Figure 3a).

Similarly, no significant grain coarsening or grain refinement has been observed for the t-HPT-10 state in Figure 12b. Nevertheless, in both cases, the crack path is influenced by the microstructure as it (partly) deviates from a straight crack front when crossing grain boundaries.

3.4. Discussion

The present investigation of CG and UFG medium carbon steels shows a correlation between hardness and endurance limit in bending fatigue tests. Four material states with ultrafine grained microstructure (s-HPT-6, s-HPT-10, t-HPT-6, and t-HPT-10), produced from two different initial carbide morphologies (s-ini + t-ini), were investigated. The refinement of the microstructure during HPT was in accordance with hardness measurements. Both s-ini and t-ini samples revealed a finer grain size and higher hardness with increasing shear strain due to torsional deformation. However, for the samples prepared by six rotations of HPT, the increase in hardness or grain refinement is not as high as expected from the applied shear strain. Nevertheless, it cannot be excluded that the anvils might have been partially sliding over the specimen during HPT deformation, which would explain the limited hardness increase between the initial states and the HPT-6 states.

Figure 14 indicates the relationship between hardness and endurance limit in our investigations. It was created on the basis of the endurance limits presented in Figure 6. A linear correlation can be seen for the hardness range up to about 500 HV. The fatigue limit primarily correlates with the hardness which itself is strongly affected by the microstructure such as the carbide and grain morphologies. The initial and HPT-6 states fit this correlation very well. Following the investigation by Murakami [26], there must be a decrease of the fatigue limit with increasing hardness values over about 500 HV. McGreevy analyzed this behavior in [27] considering the competing roles of microstructure and flaw size.

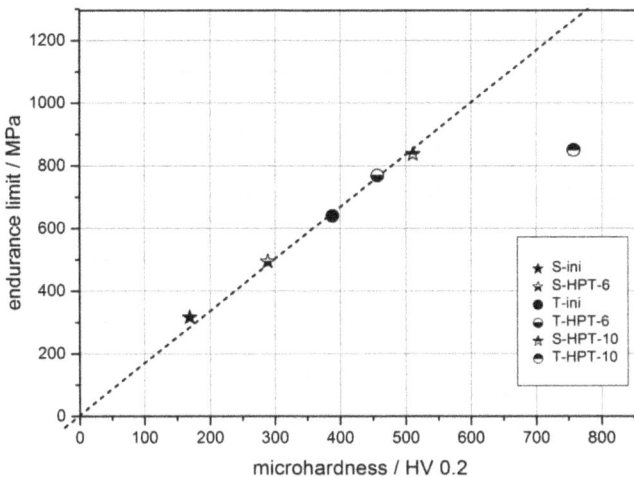

Figure 14. Relationship between hardness and endurance limit revealing a linear correspondence.

Responsible for a deviation from the linear relationship are inherent material flaws such as non-metallic inclusions, which become more and more dominant at higher hardness. The subsequent decrease of the endurance limit, as suggested in Figure 14, depends on the material state and processing route. In high strength steels, the so-called process flaws resulting from the respective treatment are responsible for such behavior. For conventional steels, it is mostly the heat treatment which leads to process flaws. In our case, the HPT treatment replaces heat treatment for reaching a high strength. Thus, instead of the traditional process, flaws known from the literature, such as micro-cracks or weakened former austenite grain boundaries in carburized steels [28], here we are confronted with shear bands or cracks created during HPT, which act as process flaws. Other possible process flaws could not be clearly identified in the metallographic sections after the fatigue testing. Nevertheless, one result of the present investigation is that, in contrast to a material that only exhibits inherent or intrinsic flaws, here the process flaws in the material state with the highest hardness (t-HPT-10) lower the fatigue limit. The S-N curves already provide evidence for this, as contrary to our expectation, no non-metallic inclusions were responsible for crack initiation in the t-HPT-10 state, but the cracks mostly initiated from the surface. This is untypical for high strength steels and indicates that there must be something inside the microstructure that is more inhomogeneous or detrimental than the non-metallic inclusions.

A more detailed investigation of the fracture surfaces confirms that the s-HPT-state exhibits a very homogenous and typical fracture appearance for high strength steels with flat fatigue fracture surfaces. The dimples that are present on the residual fracture surface indicate a ductile fracture behavior. This is known from ultrafine-grained materials and has been linked to good fatigue properties [29].

Comparing Figures 8 and 11, the fracture surfaces of t-HPT-10 do not appear as uniform as those of s-HPT-10. The main difference is that while the fatigue fracture surface in Figure 8 is not affected by inhomogeneities, Figure 11 reveals a strong topology with fracture planes of different directions covering the whole surface, including the fatigue fracture surface. In both cases shown in Figure 11, the residual fracture surface has shear steps and no dimples indicating a brittle fracture behavior. In this context, Figure 10 provides an opportunity to understand the morphology of the shear bands prior to fatigue testing. Figure 10a shows an FIB cross-section of one of these cavities, which exhibits the typical fracture appearance of the shear bands [25]. There is a narrow plane-shaped inhomogeneity beneath the surface inside the microstructure. This could be a link to shear bands generated during HPT as similar stripes are present throughout the microstructure before fatigue loading, as visible in the optical micrograph in Figure 10c. This observation can be made not only in the brittle tempered state but similarly also in the ductile spheroidizing annealed microstructure with excellent fatigue properties. It follows that s-HPT-10 and t-HPT-10 do contain shear bands prior to fatigue testing but only in the tempered state they affect the fatigue properties. Therefore, the brittle behavior is the main reason for fatigue limits which are lower than expected in consideration of the hardness, because the brittle material has a more limited ability to reduce stress around process flaws owing to plastic flow. In this context, the s-HPT-10 state is an ideal material in respect of fatigue because it exhibits a high hardness and also a high ductility resulting in excellent fatigue performance with higher tolerance against crack initiation at inhomogeneities.

However, an open question is: why do early cracks or cracks at low stress amplitudes occur in some specimens of t-HPT-10? This can be explained comparing Figures 8 and 11, which both show shear bands in the fracture surface. Shear bands represent inhomogeneities, which, in combination with a brittle material, are detrimental to the fatigue performance, as shown in Figure 14. However, when comparing fracture surfaces of late and early failure, it is visible that the specimens with the worst fatigue properties exhibit a rotation of the shear bands, which is not parallel to the nominal HPT-shear planes. Unusually turbulent flows, manifesting themselves in the appearance of whirls observed at the macroscopic [30] and microscopic [31] scale in severely deformed materials, are most likely the reason for this rotation. This effect seems to increase the influence of the shear bands on fatigue failure.

The rotation of the shear bands increases the possibility for cracks to grow along the shear bands. Investigations from Miller [32,33] support this argumentation. He proposed that cracks are always present with every kind of inhomogeneity and the only determining factor for the fatigue resistance is the question whether they reach a critical length that gives them the possibility of further growing. In the case of ultrafine grained high strength steels, grain boundaries, which are the traditional microstructural barrier against crack growth, are not efficient anymore because of the size of the inherent flaws such as non-metallic inclusions. In contrast to the grain size the inclusions do not decrease in size with increasing hardness. The inclusions can be regarded as cracks that are larger than the grain size and cannot be stopped easily by a single grain boundary. Process flaws, such as the identified shear bands, are, depending on their appearance, more detrimental than the non-metallic inclusions. However, important for reaching a critical length is not only the absolute length of a shear band. Crucial is also their direction in relation to the maximum applied stress direction [26,27]. With this argument it becomes clear that the rotated shear bands are more critical than those in the HPT planes. The flat fatigue crack path along the shear planes also indicates a higher fatigue crack growth rate, which lowers the lifetime of the samples, as shown in the S-N diagrams in Figure 6. This observation is known from the literature as well as the fact that the threshold for long crack propagation is lower for favorably oriented shear bands [34]. Evidence for this explanation and for the crack initiation being affected by the shear bands and their orientation is presented in Figure 12, where the shear bands are visible on the fracture surface and also at the lower side of the specimen with the highest fatigue stresses. This characteristic appearance has already been reported in literature [35] and seems to be beneficial for crack initiation.

4. Conclusions

The present investigation offers an analysis of the fatigue behavior of medium carbon steel SAE 1045 with different initial carbide morphologies (spheriodized and tempered state) processed by HPT. High hardness values of up to 758 HV and a homogenous ultrafine grained microstructure were observed. Microstructure investigations and the results of bending fatigue tests led to the following conclusions:

- The microstructure was refined during HPT. The ferrite grain refinement during HPT is more pronounced for the state with fine dispersed carbides obtained by tempering as compared to the state with spheroidal carbides. Six and ten rotations were used for HPT processing for each of the heat treatment conditions. Only after ten rotations the microstructure was fine enough to reach highest strength and hardness.

- Up until circa 500 HV or approximately 830 MPa the endurance limits correlate linearly with the hardness. The carbide morphology does not affect the linear behavior of the fatigue limit directly in this hardness region. In addition, no process flaws or other inherent flaws significantly influenced the fatigue behavior except for the hardness. The reason is the high ductility of the spheroidizing annealed microstructure, which is tolerant towards stress concentration at shear bands.

- Above approximately 500 HV, the fatigue limit no longer correlates linearly with the hardness. Process flaws are the main reason for this behavior. The process flaws observed in the present investigation are shear bands caused by the HPT treatment. Evidence was found that shear bands were responsible for crack initiation at low fatigue loads and high crack propagation rates. This is particularly noticeable when the shear bands are rotated out of the plane of the fatigue specimen.

Acknowledgments: The authors would like to thank the German Research Foundation (DFG) for the financial support of this work under grant IV 98/4-1 and KE 1426/3-1. We acknowledge support by Open Access Publishing Fund of Karlsruhe Institute of Technology.

Author Contributions: Y.I. processed samples using HPT. C.R. conducted fatigue tests and performed the metallographic and fractographic observations. A.K. and E.C.-M. conducted characterization of the microstructure by TEM. R.P. prepared the TEM samples using FIB. C.R. performed analysis of the data. E.K., Y.I. and C.K. supervised the work. C.R. and E.K. wrote the initial manuscript with input from all authors. All authors contributed to discussion of the results, provided input on the manuscript, and approved the final version.

References

1. Song, R.; Ponge, D.; Raabe, D.; Speer, J.G.; Matlock, D.K. Overview of processing, microstructure and mechanical properties of ultrafine grained bcc steels. *Mater. Sci. Eng. A* **2006**, *441*, 1–17.
2. Shen, H.; Li, Z.; Gunther, B.; Korznikov, A.V.; Valiev, R.Z. Influence of powder consolidation methods on the structural and thermal properties of a nanophase Cu-50wt%Ag alloy. *Nanostructured Mater.* **1995**, *6*, 385–388.

3.   Valiev, R.Z.; Langdon, T.G. Principles of equal-channel angular pressing as a processing tool for grain refinement. *Prog. Mater. Sci.* **2006**, *51*, 881–981.

4.   Zhilyaev, A.P.; Langdon, T.G. Using high-pressure torsion for metal processing: Fundamentals and applications. *Prog. Mater. Sci.* **2008**, *53*, 893–979.

5.   Tsuji, N. Fabrication of bulk nanostructured materials by Accumulative Roll Bonding (ARB). In *Bulk Nanostructured Materials*; Zehetbauer, M.J., Zhu, Y.T., Eds.; Wiley-VCH: Weinheim, Germany, 2009.

6.   Valiev, R.Z.; Islamgaliev, R.K.; Alexandrov, I.V. Bulk nanostructured materials from severe plastic deformation. *Prog. Mater. Sci.* **2000**, *45*, 103–189.

7.   Pouryazdan, M.; Schwen, D.; Wang, D.; Scherer, T.; Hahn, H.; Averback, R.S.; Bellon, P. Forced chemical mixing of immiscible Ag-Cu heterointerfaces using high-pressure torsion. *Phys. Rev. B* **2012**, *86*, 144302.

8.   Korznikov, A.V.; Dimitrov, O.; Korznikova, G.F.; Dallas, J.P.; Quivy, A.; Valiev, R.Z.; Mukherjee, A. Nanocrystalline structure and phase transformation of the intermetallic compound TiAl processed by severe plastic deformation. *Nanostructured Mater.* **1999**, *11*, 17–23.

9.   Pavlov, V.A. Structural amorphization of metals and alloys with an extremely high plastic deformation ratio. *Phys. Metals Metallogr.* **1985**, *59*, 629–649.

10. Valiev, R. Nanostructuring of metals by severe plastic deformation for advanced properties. *Nat. Mater.* **2004**, *3*, 511–516.

11. Valiev, R.Z.; Murashkin, M.Y.; Kilmametov, A.; Straumal, B.; Chinh, N.Q.; Langdon, T.G. Unusual super-ductility at room temperature in an ultrafine-grained aluminum alloy. *J. Mater. Sci.* **2010**, *45*, 4718–4724.

12. Vinogradov, A.; Patlan, V.; Suzuki, Y.; Kitagawa, K.; Kopylov, V.I. Structure and properties of ultra-fine grain Cu–Cr–Zr alloy produced by equal-channel angular pressing. *Acta Mater.* **2002**, *50*, 1639–1651.

13. Vinogradov, A.; Hashimoto, S. Multiscale Phenomena in Fatigue of Ultra-Fine Grain Materials—An Overview. *Mater. Trans.* **2001**, *42*, 74–84.

14. Chapetti, M.D.; Miyata, H.; Tagawa, T.; Miyata, T.; Fujioka, M. Fatigue strength of ultra-fine grained steels. *Mater. Sci. Eng. A* **2004**, *381*, 331–336.

15. Sawai, T.; Matsuoka, S.; Tsuzaki, K. Low- and High-cycle Fatigue Properties of Ultrafine-grained Low Carbon Steels. *Tetsu-to-Hagane* **2003**, *89*, 726–733.

16. Niendorf, T.; Canadinc, D.; Maier, H.J.; Karaman, I.; Sutter, S.G. On the fatigue behavior of ultrafine-grained interstitial-free steel. *Int. J. Mat. Res.* **2006**, *97*, 1328–1336.

17. Okayasu, M.; Sato, K.; Mizuno, M.; Hwang, D.Y.; Shin, D.H. Fatigue properties of ultra-fine grained dual phase ferrite/martensite low carbon steel. *Int. J. Fatigue* **2008**, *30*, 1358–1365.

18. Niendorf, T.; Böhner, A.; Höppel, H.W.; Göken, M.; Valiev, R.Z.; Maier, H.J. Comparison of the monotonic and cyclic mechanical properties of ultrafine-grained low carbon steels processed by continuous and conventional equal channel angular pressing. *Mater. Des.* **2013**, *47*, 138–142.

19. Ning, J.-L.; Courtois-Manara, E.; Kurmanaeva, L.; Ganeev, A.V.; Valiev, R.Z.; Kübel, C.; Ivanisenko, Y. Tensile properties and work hardening behaviors of ultrafine grained carbon steel and pure iron processed by warm high pressure torsion. *Mater. Sci. Eng. A* **2013**, *581*, 8–15.

20. Hück, M. Ein verbessertes Verfahren für die Auswertung von Treppenstufenversuchen. *Z. Werkstofftech.* **1983**, *14*, 406–417.

21. Rauch, E.F.; Portillo, J.; Nicolopoulos, S.; Bultreys, D.; Rouvimov, S.; Moeck, P. Automated nanocrystal orientation and phase mapping in the transmission electron microscope on the basis of precession electron diffraction. *Zeitschrift für Kristallographie* **2010**, *225*, 103–109.

22. Bachmann, F.; Hielscher, R.; Schaeben, H. Grain detection from 2d and 3d EBSD data—Specification of the MTEX algorithm. *Ultramicroscopy* **2011**, *111*, 1720–1733.

23. Kobler, A.; Kashiwar, A.; Hahn, H.; Kübel, C. Combination of in situ straining and ACOM TEM: A novel method for analysis of plastic deformation of nanocrystalline metals. *Ultramicroscopy* **2013**, *128*, 68–81.

24. Kobler, A. Untersuchung von Deformationsmechanismen in nanostrukturierten Metallen und Legierungen mit Transmissionselektronenmikroskopie. Ph.D. Thesis, Technical University Darmstadt, TUPrints, Darmstadt, Germany, 2015.

25. Ivanisenko, Y.; Kurmanaeva, L.; Weissmueller, J.; Yang, K.; Markmann, J.; Rösner, H.; Scherer, T.; Fecht, H.-J. Deformation mechanisms in nanocrystalline palladium at large strains. *Acta Mater.* **2009**, *57*, 3391–3401.

26. Murakami, Y.; Kodama, S.; Konuma, S. Quantitative evaluation of effects of non-metallic inclusions on fatigue strength of high strength steels. *Int. J. Fatigue* **1989**, *11*, 291–298.

27. McGreevy, T.E.; Socie, D.F. Competing roles of microstructure and flaw size. *Fatigue Fract. Engng. Mater. Struct.* **1999**, *22*, 495–508.

28. Apple, C.A.; Krauss, G. Microcracking and Fatigue in a Carburized Steel. *Metall. Trans.* **1973**, *4*, 1195–1200.

29. Suresh, S. *Fatigue of Materials*; Cambridge Solid State Science Series; Cambridge University Press: Cambridge, UK, 1998.

30. Cao, Y.; Kawasaki, M.; Wang, Y.B.; Alhajeri, S.N.; Liao, X.Z.; Zheng, W.L.; Ringer, S.P.; Zhu, Y.T.; Langdon, T.G. Unusual macroscopic shearing patterns observed in metals processed by high-pressure torsion. *J. Mater. Sci.* **2010**, *45*, 4545–4553.

31. Ivanisenko, Y.; Lojkowski, W.; Valiev, R.Z.; Fecht, H.-J. The mechanism of formation of nanostructure and dissolution of cementite in a pearlitic steel during high pressure torsion. *Acta Mater.* **2003**, *51*, 5555–5570.

32. Miller, K.J. Materials science perspective of metal fatigue resistance. *Anales de Mecanica de la Fractura* **1995**, *12*, 1–10.

33. Miller, K.J. The two thresholds of fatigue behaviour. *Fatigue Fract. Engng. Mater. Struct.* **1993**, *16*, 931–939.

34. Hockauf, K.; Hockauf, M.; Wagner, M.F.-X.; Lampke, T.; Halle, T. Fatigue crack propagation in an ECAP-processed aluminium alloy—Influence of shear plane orientation. *Materialwissenschaft und Werkstofftechnik* **2012**, *43*, 609–616.

35. Hamada, A.S. Deformation and Damage Mechanisms in Ultrafine-Grained Austenitic Stainless Steel During Cyclic Straining. *Metall. Mater. Trans. A* **2013**, *44*, 1626–1630. [CrossRef]

![metals logo] *metals*

MDPI

Communication

Mechanical Behavior of Ultrafine Gradient Grain Structures Produced via Ambient and Cryogenic Surface Mechanical Attrition Treatment in Iron

Heather A. Murdoch *, Kristopher A. Darling, Anthony J. Roberts and Laszlo Kecskes

U.S. Army Research Laboratory, Weapons and Materials Directorate, Aberdeen Proving Ground, MD 21005-5069, USA; kristopher.a.darling.civ@mail.mil (K.A.D.); anthony.j.roberts69.ctr@mail.mil (A.J.R.); laszlo.j.kecskes.civ@mail.mil (L.K.)

* Author to whom correspondence should be addressed; heather.a.murdoch.civ@mail.mil; Tel.: +1-410-306-0699; Fax: +1-410-306-0759.

Academic Editor: Heinz Werner Höppel

Received: 31 March 2015; Accepted: 21 May 2015; Published: 3 June 2015

Abstract: Ambient and cryogenic surface mechanical attrition treatments (SMAT) are applied to bcc iron plate. Both processes result in significant surface grain refinement down to the ultrafine-grained regime; the cryogenic treatment results in a 45% greater grain size reduction. However, the refined region is shallower in the cryogenic SMAT process. The tensile ductility of the grain size gradient remains low (<10%), in line with the expected behavior of the refined surface grains. Good tensile ductility in a grain size gradient requires the continuation of the gradient into an undeformed region.

Keywords: grain size gradient; surface mechanical attrition treatment; cryogenic; ultrafine-grained

1. Introduction

Numerous reports now exist indicating an order of magnitude increase in strength is possible in metals and alloys that exhibit grain sizes approaching the lower limit of nanocrystallinity. While achieving high strength has never been a problem, the ability to achieve any amount of uniform elongation (the prerequisite for appreciable ductility) has been a challenge. However, several methods have recently been developed to mitigate this strength-ductility tradeoff through the engineering of multi-length scale structures including bimodal grain size distributions [1,2], nanoscale twins [3,4], and grain size gradients [5]. Specifically, gradient microstructures generated through surface mechanical attrition treatments or SMAT have additional benefits over other hierarchical microstructures from a surface science/tribological standpoint by concentrating the nanocrystalline properties in the surface region. For instance, nanostructured surface layers have shown improved corrosion resistance [6–9], wear [10,11] and fatigue [12–14], and irradiation resistance [15].

Current SMAT techniques have shown to be very efficient methods for producing grain size gradients, inducing substantial surface grain refinement and varying depths and grades of grain refinement. It has been shown that differences in processing methods can greatly affect both the overall structures (e.g., depth of refined region and "slope" of the grain size gradient) and the individual microstructures (e.g., surface grain size [16], deformation artifacts within grain size regions [16,17]). It was noted by Tao *et al.* [17] in their work introducing SMAT that finer grains would be expected with plastic deformation at lower temperatures. Indeed, Darling *et al.* provided the first evidence in a brief report for this effect through a cryogenic SMAT process on copper [16]. The percent reduction in grain size (60%) due to cryogenic processing is in good agreement with the empirical correlation between the resulting grain size and the Zener-Holloman parameter (combined metric of strain rate and deformation temperature) [18]. In a magnesium alloy, a different surface treatment method resulted in a 63% decrease in grain size for cryogenic burnishing *versus* ambient [19].

As compared to copper, bcc iron would be expected to have a lesser reduction in grain size based on this empirical parameter in addition to the differences in plastic deformation behavior; nanocrystalline/ultrafine-grained iron exhibits essentially no strain hardening [20,21] and an inverse relationship with strain rate sensitivity as compared to fcc materials [21]—especially important as the strain rates involved in the SMAT process are relatively high ($\sim 10^2$). The cryogenic SMAT process will also take the iron well below its ductile to brittle transition temperature. In this work, we look at the effects of cryogenic and ambient SMAT processing on the microstructure and mechanical properties of iron.

2. Experimental Section

The SMAT process was applied to 0.6 cm thick discs 6.35 cm in diameter cut from a rod of ARMCO iron (Goodfellow, Huntington, UK; purity > 99.85%Fe). Details of both the cryogenic and ambient SMAT processes can be found in [16]. Briefly, the material to be treated is fitted onto one end of the vial in a mechanical alloying mill (SPEX, Company, Metuchen, NJ, USA); the milling media within the vial, in this case 50 g of stainless steel shot, continually impacts the surface at high rate and variable direction during the SMAT process. For the cryogenic SMAT process, the milling vial is enclosed by a Teflon sleeve through which liquid nitrogen is continuously flowing throughout the treatment. The iron plates were polished to a mirror finish before treatment. The SMAT process was performed for one hour for both the ambient and cryogenic treatments.

Following the SMAT processes, the plates were sectioned and polished by a series of steps down to 1 μm alumina. The microstructural analysis was performed using an FEI Nova 600i dual beam (FEI, Hillsboro, OR, USA) Focused Ion Beam (FIB) system. Focused ion beam channeling contrast images (FIBCCI) are obtained using backscattered electrons produced by the ion beam as it rasters across the sample surface. The FIBCCI contrast mechanism is due to changes in the grain orientations that cause variations in ion channeling efficiency, *i.e.*, crystals which are able to channel more effectively due to their orientation produce fewer detectable electrons, so orientations closer to incident ions show up darker, *i.e.*, crystal orientation specific contrast.

Hardness measurements were obtained with a Wilson Hardness Tukon 1202 (Buehler, Lake Bluff, IL, USA) using a load of 50 g load with 10 s dwell time with three measurements at each depth. Tensile test dogbones were cut from the SMAT plates with a MicroProtoSystems DSLS 3000 micromill (MicroProtoSystems, Chandler, AZ, USA) with the approximate gauge dimensions: 5 mm length, 1 mm width, and thickness of ~350 μm. The tensile tests were performed on a custom miniature tensile test apparatus which utilizes digital image correlation to track the sample extension. Three tensile tests for each sample were performed at a load rate of 2 μm/s with a 125 lb load cell.

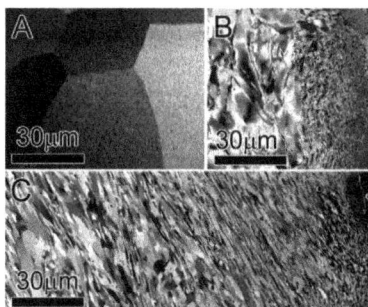

Figure 1. (**a**) Initial microstructure of the ARMCO Iron plate; (**b**) Surface microstructure following the cryogenic surface mechanical attrition treatment (SMAT), showing considerable grain refinement and plastic deformation; (**c**) Surface microstructure following ambient SMAT treatment. The grain refinement continues a considerable distance into the material.

3. Results and Discussion

3.1. Microstructure

The FIBCCI contrast micrograph (Figure 1A) reveals the initial grain size of the iron plate to be 50–100 μm. After the SMAT treatment at ambient temperature (Figure 1C), the plate exhibited submicron grains up to ~200 μm deep into the sample, with plastic deformation artifacts continuing up to about 700 μm. The average surface grain size (measured within the top 5 μm of the plate) was 650 nm. In contrast, the average surface grain size for the cryogenic SMAT treatment of the same duration was 350 nm (Figure 1C). As in the case of cryogenic SMAT copper, which showed a 60% reduction in grain size with respect to the ambient [16], the cryogenic SMAT iron followed the same trend of higher grain refinement than the ambient SMAT treatment, but to a lesser extent. The reduction of grain size by only ~45% in the iron follows literature trends for microstructural refinement as described by the strain/temperature pairing through the Zener-Holloman parameter [18,22]. Iron has higher activation energy for deformation than copper, generally taken in pure metals as similar to the activation energy for self-diffusion; therefore, the grain refinement is less receptive to changes in temperature. In addition to the differences in surface grain size, the grain size gradient in the cryogenic SMAT iron is significantly sharper, exhibiting only a ~50 μm region of submicron grains and ~300 μm region of plastic deformation. The surface of the cryogenic SMAT iron also shows some surface cracks, as can be seen in the far upper right of Figure 1B.

3.2. Mechanical Properties

The microhardness as a function of depth into the plate is shown in Figure 2. The cryogenic SMAT sample had a higher surface hardness of 2.6 GPa compared to the ambient SMAT plate of 2.4 GPa, in line with predictions from the Hall-Petch relationship for iron [23]. The hardness of the cryogenic SMAT plate reduces more rapidly than in the ambient SMAT plate—dropping from 2.6 GPa to 2 GPa within the first 50 μm and then to ~1.7 GPa within the first 100 μm, mirroring the steepness of the gradient compared to the ambient cross sections. However, after the first ~100 μm, there is no significant difference in the hardness—as the grain size increases out of the ultrafinegrained regime, the variance in hardness with changes in grain size is minimal. Additionally, while the grain size grows rapidly in the cryogenic SMAT plate, the larger grains still contain a significant amount of deformation artifacts such as dislocation walls and tangles [17,24,25], as can be seen in the changing contrast in the channeling images. These microstructural features, internal to the grain boundaries, can also contribute to the observed hardness of the material.

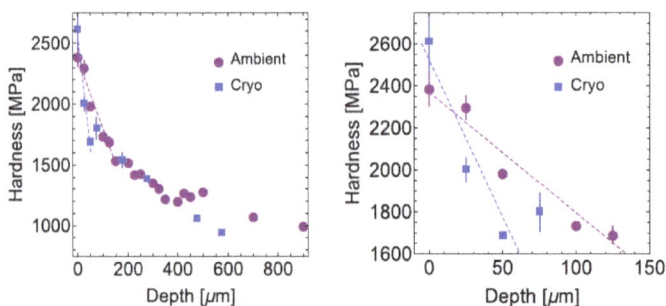

Figure 2. Microhardness of iron plates treated by the SMAT process as a function of depth into the sample. The values for the cryogenic SMAT iron are indicated by blue squares; ambient SMAT by purple circles. Dashed lines are a guide to the eye.

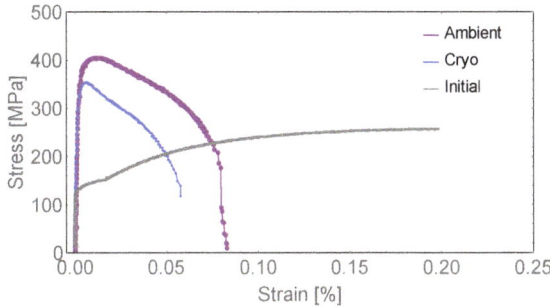

Figure 3. Tensile behavior for cryogenic SMAT (blue curve), ambient SMAT (purple curve) iron, and untreated iron plate (gray curve). The cryogenic SMAT exhibits lower strength and ductility than the ambient SMAT. Both SMAT treated plates show improvement in strength and decrease in ductility as compared to the untreated iron.

The yield strengths of the cryogenic and ambient SMAT iron samples were 345 and 385 MPa respectively, significantly higher than that of the untreated iron plate (~150 MPa), as seen in Figure 3. While the higher surface grain refinement in the cryogenic SMAT gradient led to a higher surface hardness commensurate with Hall-Petch behavior (Figure 2), the opposite relationship is observed in the tensile tests, with the ambient SMAT iron exhibiting a higher yield strength. The gradient structure in the cryogenic SMAT plate only exhibited significant grain refinement to a depth of about 50 μm, comprising about 14% of the tensile dogbone thickness. The ambient SMAT gradient penetrated much deeper into the plate, encompassing closer to 60% of the thickness of the tensile specimen. A greater volume fraction of the tensile specimen is then comprised of ultrafine grains in the ambient SMAT iron, resulting in a higher overall yield strength. A similar result was observed in tensile specimens composed of a gradient twin layers, wherein the depth of the gradient into the dogbone sample was found to correspond to the strength according to a rule of mixtures [26]. An additional contribution to the poor mechanical behavior in the cryogenic SMAT iron may be due to the difference in surface condition between the two processes. As can be seen in Figure 1B, the surface of the cryogenic SMAT plate can exhibit small cracks than are attributed to the expected brittle (*versus* ductile) behavior at the greatly reduced processing temperature.

Both the cryogenic and ambient SMAT iron displayed very little uniform elongation before exhibiting significant strain softening in contrast to the strain hardening behavior of the initial iron plate (Figure 3). Nanocrystalline bcc iron usually exhibits brittle fracture in tension, while strain softening is observed in ultrafinegrained iron [20,27,28] in grain sizes as large as 4 μm [29]. A stress drop (e.g., the amount of softening) of ~400 MPa from yield to failure has been observed in homogenous ultrafinegrained iron samples of similar grain size to that of the surface grains in this work [20,28–30]. The total elongation is also in line with homogenous ultrafinegrained iron produced through ECAP [20,28,30] while the overall strength is lower due to the increasing grain size of the gradient out of the ultrafinegrained regime.

In contrast to this observed strain softening behavior, it was reported in [25] that a grain size gradient structure in steel exhibited extraordinary strain hardening; additionally, a grain size gradient in copper displayed significant strain hardening and tensile elongation as well [5]. To examine these differences, we first look at the two literature reports of significant strain hardening in grain size gradients. In contrast to iron, nanocrystalline and ultrafine-grain copper can display some strain hardening behavior [31]. Additionally, in the case of the grain size gradient in [5], the significant plastic deformation is found to be dominated by mechanically driven grain growth throughout the grain size gradient during loading.

The strain hardening behavior in the steel grain size gradient was not found to be a result of mechanical grain growth [25]; the varying regions of the gradient structure were tested separately and together to reveal a synergistic effect between the gradient region and the undeformed core material. When the gradient structure was isolated (e.g., the top 120 μm of the sample was tested separately), it did not exhibit strain hardening behavior, rather strain softening, with a yield drop of almost 100 MPa and elongation of <10%. Significant strain hardening was observed only when the tensile sample thickness included both the gradient layer and a considerable fraction of the undeformed steel; the gradient layer represented about 12% of the tensile sample thickness [25]. Additionally, the hardening behavior was measured as a function of depth through hardness measurements that were performed after the tensile test. Hardening was only exhibited towards the back end of the gradient structure (as the gradient transition to the undeformed core), where the grain size was much greater than 1 μm.

In this work on grain size gradient iron, the thickness of the tensile test samples was ~350 μm which encompasses only the refined grain size gradient and heavily deformed regions, and none of the pristine non-deformed coarse-grained core. The low tensile ductility is therefore a result of this truncation of the gradient before reaching the undeformed core, the tensile behavior of both the cryogenic and ambient SMAT processed iron is in line with that of the stand-alone steel gradient layer in [25] which exhibited low elongation and a lack of strain hardening. Standalone tests of the gradient surface layer in copper were also consistent with this mechanical behavior [5,32].

While still exhibiting good ductility, the grain size gradient in Cu-Zn [33] does not improve upon the strength-ductility tradeoff in homogenous grain size materials as significantly as the copper [5] and steel [25] gradients. The grain refinement in the Cu-Zn study is not quantified, but most of the grain sizes in the gradient appear to be much larger than 1 μm; additionally, the hardness measurements indicate that the entire thickness of the tensile samples (600 μm) has been plastically deformed, preventing the unusual elongation and hardening behavior accessed by the studies including the undeformed core [5,25].

The standard strength-ductility tradeoff associated with grain refinement is shown in Figure 4—a typical boundary region is shown with the dashed curve drawn through data points for homogeneous grain structures of the same material as the gradient structures: pure iron [34] (gray circles), pure copper [31,35–40] (gray diamonds), steel [25] (gray triangles), and Cu-Zn alloys [33] (gray squares). These points are data from bulk samples of various homogeneous grain sizes and processing methods for comparison with the gradient grain structures of the sample material. The strength and elongation of the existing data for grain size gradient structures – this work in iron (magenta circles), Cu-Zn alloys (red squares), steel (black triangle), and copper (orange diamond) are plotted with respect to the bulk literature data. Only the gradient structures that include a significant fraction of non-deformed grains in the tensile specimen (copper [5] and steel [25]) are significantly off the tradeoff curve for their pure homogenous counterparts. This further supports the work of [25], which describes the unusual synergistic effect of the deformed gradient layer and the coarse-grained core. The typical strain softening behavior in bcc iron and the lack of an undeformed core section in the tensile specimens in this work explain the relatively poor position on the frontier of the strength-ductility tradeoff for the cryogenic and ambient SMAT iron, as compared to the other three gradient systems.

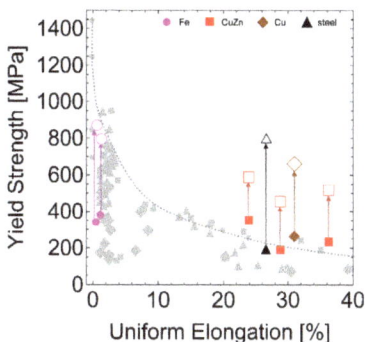

Figure 4. Tensile data existing for gradient grain structures are depicted for the iron SMAT in this work (magenta circles), Cu-Zn alloys [33] (red squares), copper [5] (orange diamond); steel [25] (black triangle). Literature values for bulk structures in the same materials are shown in gray (iron: gray circles [34], steel: grey triangles [25] and references therein, Cu-Zn alloys: grey squares [33] and references therein, copper: gray diamonds [31,35–40]). The strength ductility tradeoff is illustrated by the dotted line. The surface strength for each gradient structure is depicted with an open symbol of the same color and at the same elongation point, connected by a gradient arrow.

The surface strength of the grain size gradient structures is also plotted in Figure 4, at the sample elongation as the gradient material and marked with an open symbol of the same color. The surface yield strength shown is calculated from surface hardness measurements (yield ~H/3) for the iron in this work, the Cu-Zn example, and the steel example; the surface yield strength for the pure copper example was from a tensile test of a free standing foil cut from the surface. In the work on steel [25], strength measurements reported from tensile tests performed with a foil cut from the surface layer and the hardness tests were congruent. While the overall strength and ductility of the gradient structure may not be a significant improvement on a bulk sample of a similar grain size, the surface strength is consistently a marked improvement over a homogenous grain size structure at the same elongation. This difference highlights an engineering advantage of grain size gradient materials, in that the surface of a grain size gradient structure can be as much as eight times harder than a homogenous grain size part of similar ductility.

4. Conclusions

The application of surface mechanical attrition treatment at both ambient and cryogenic temperatures to bcc iron plate resulted in significant surface grain refinement and resulted in a grain size gradient. The cryogenic SMAT produced a 45% greater grain size reduction as compared to the ambient SMAT, but a shallower depth of grain refinement. Consequently, the surface hardness was higher for the cryogenic SMAT, but the tensile strength and ductility was lower, due to the lower volume fraction of ultrafine-grains. Strain softening is observed, in line with iron with homogenous grain sizes in the ultrafine-grain regime. The tensile elongation of both grain size gradients remains low (<10%), in contrast to the extraordinary strain hardening observed in grain size gradient work in steel [25], due to the lack of undeformed core region in the tensile samples. Moving forward, the relationship between the volume fraction of gradient grains/deformed region and ductility should be explored in order to successfully exploit the benefits of nanocrystalline surface layers while maintaining ductility in the larger part.

Acknowledgments: The authors would like to acknowledge Jim Catalano (ARL), Tom Luckenbaugh (Bowhead) and Micah Gallagher (Bowhead) for valuable experimental input. AJR would like to acknowledge support from the U.S. Army Research Laboratory administered by the Oak Ridge Institute for Science and Education through an interagency agreement between the U.S. Department of Energy and ARL.

Author Contributions: H.A.M. and K.A.D. conceived and designed the experiments; A.J.R. and H.A.M performed the experiments; H.A.M. and K.A.D. analyzed the data; H.A.M. created the initial draft and all authors discussed the results, participated in the manuscript preparation, and approved the final manuscript.

Conflicts of Interest: The authors declare no conflict of interest.

References

1. Wang, Y.; Chen, M.; Zhou, F.; Ma, E. High tensile ductility in a nanostructured metal. *Nature* **2002**, *419*, 912–915. [CrossRef] [PubMed]
2. Srinivasarao, B.; Oh-ishi, K.; Ohkubo, T.; Hono, K. Bimodally grained high-strength Fe fabricated by mechanical alloying and spark plasma sintering. *Acta Mater.* **2009**, *57*, 3277–3286. [CrossRef]
3. Lu, L.; Chen, X.; Huang, X.; Lu, K. Revealing the maximum strength in nanotwinned copper. *Science* **2009**, *323*, 607–610. [CrossRef] [PubMed]
4. You, Z.S.; Lu, L.; Lu, K. Tensile behavior of columnar grained Cu with preferentially oriented nanoscale twins. *Acta Mater.* **2011**, *59*, 6927–6937. [CrossRef]
5. Fang, T.H.; Li, W.L.; Tao, N.R.; Lu, K. Revealing extraordinary intrinsic tensile plasticity in gradient nano-grained copper. *Science* **2011**, *331*, 1587–1590. [CrossRef] [PubMed]
6. Balusamy, T.; Sankara Narayanan, T.S.N.; Ravichandran, K.; Park, I.S.; Lee, M.H. Influence of surface mechanical attrition treatment (SMAT) on the corrosion behaviour of aisi 304 stainless steel. *Corros. Sci.* **2013**, *74*, 332–344. [CrossRef]
7. Fabijanic, D.; Taylor, A.; Ralston, K.D.; Zhang, M.X.; Birbilis, N. Influence of surface mechanical attrition treatment attrition media on the surface contamination and corrosion of magnesium. *Corrosion* **2012**, *69*, 527–535. [CrossRef]
8. Huang, R.; Han, Y. The effect of smat-induced grain refinement and dislocations on the corrosion behavior of Ti–25Nb–3Mo–3Zr–2Sn alloy. *Mater. Sci. Eng C* **2013**, *33*, 2353–2359. [CrossRef] [PubMed]
9. op't Hoog, C.; Birbilis, N.; Estrin, Y. Corrosion of pure Mg as a function of grain size and processing route. *Adv. Eng. Mater.* **2008**, *10*, 579–582. [CrossRef]
10. Dai, K.; Shaw, L. Comparison between shot peening and surface nanocrystallization and hardening processes. *Mater. Sci. Eng. A* **2007**, *463*, 46–53. [CrossRef]
11. Ortiz, A.L.; Tian, J.-W.; Shaw, L.L.; Liaw, P.K. Experimental study of the microstructure and stress state of shot peened and surface mechanical attrition treated nickel alloys. *Scr. Mater.* **2010**, *62*, 129–132. [CrossRef]
12. Shaw, L.L.; Tian, J.-W.; Ortiz, A.L.; Dai, K.; Villegas, J.C.; Liaw, P.K.; Ren, R.; Klarstrom, D.L. A direct comparison in the fatigue resistance enhanced by surface severe plastic deformation and shot peening in a C-2000 superalloy. *Mater. Sci. Eng. A* **2010**, *527*, 986–994. [CrossRef]
13. Villegas, J.C.; Shaw, L.L.; Dai, K.; Yuan, W.; Tian, J.; Liaw, P.K.; Klarstrom, D.L. Enhanced fatigue resistance of a nickel-based hastelloy induced by a surface nanocrystallization and hardening process. *Philos. Mag. Lett.* **2005**, *85*, 427–438. [CrossRef]
14. Roland, T.; Retraint, D.; Lu, K.; Lu, J. Fatigue life improvement through surface nanostructuring of stainless steel by means of surface mechanical attrition treatment. *Scr. Mater.* **2006**, *54*, 1949–1954. [CrossRef]
15. Efe, M.; El-Atwani, O.; Guo, Y.; Klenosky, D.R. Microstructure refinement of tungsten by surface deformation for irradiation damage resistance. *Scr. Mater.* **2014**, *70*, 31–34. [CrossRef]
16. Darling, K.A.; Tschopp, M.A.; Roberts, A.J.; Ligda, J.P.; Kecskes, L.J. Enhancing grain refinement in polycrystalline materials using surface mechanical attrition treatment at cryogenic temperatures. *Scr. Mater.* **2013**, *69*, 461–464. [CrossRef]
17. Tao, N.R.; Wang, Z.B.; Tong, W.P.; Sui, M.L.; Lu, J.; Lu, K. An investigation of surface nanocrystallization mechanism in Fe induced by surface mechanical attrition treatment. *Acta Mater.* **2002**, *50*, 4603–4616. [CrossRef]
18. Li, Y.; Zhang, Y.; Tao, N.; Lu, K. Effect of the zener–hollomon parameter on the microstructures and mechanical properties of Cu subjected to plastic deformation. *Acta Mater.* **2009**, *57*, 761–772. [CrossRef]
19. Pu, Z.; Song, G.L.; Yang, S.; Outeiro, J.C.; Dillon, O.W., Jr.; Puleo, D.A.; Jawahir, I.S. Grain refined and basal textured surface produced by burnishing for improved corrosion performance of AZ31B Mg alloy. *Corros. Sci.* **2012**, *57*, 192–201. [CrossRef]

20. Han, B.; Mohamed, F.; Lavernia, E. Mechanical properties of iron processed by severe plastic deformation. *Metall. Mater. Trans. A* **2003**, *34*, 71–83. [CrossRef]
21. Wei, Q. Strain rate effects in the ultrafine grain and nanocrystalline regimes-influence on some constitutive responses. *J. Mater. Sci.* **2007**, *42*, 1709–1727. [CrossRef]
22. Vorhauer, A.; Pippan, R. On the onset of a steady state in body-centered cubic iron during severe plastic deformation at low homologous temperatures. *Metall. Mater. Trans. A* **2008**, *39*, 417–429. [CrossRef]
23. Malow, T.R.; Koch, C.C. Mechanical properties, ductility, and grain size of nanocrystalline iron produced by mechanical attrition. *Metall. Mat. Trans. A* **1998**, *29*, 2285–2295. [CrossRef]
24. Lu, K.; Lu, J. Nanostructured surface layer on metallic materials induced by surface mechanical attrition treatment. *Mater. Sci. Eng. A* **2004**, *375–377*, 38–45. [CrossRef]
25. Wu, X.; Jiang, P.; Chen, L.; Yuan, F.; Zhu, Y.T. Extraordinary strain hardening by gradient structure. *Proc. Natl. Acad. Sci. USA* **2014**, *111*, 7197–7201. [CrossRef] [PubMed]
26. Wang, H.T.; Tao, N.R.; Lu, K. Architectured surface layer with a gradient nanotwinned structure in a Fe–Mn austenitic steel. *Scr. Mater.* **2013**, *68*, 22–27. [CrossRef]
27. Jia, D.; Ramesh, K.T.; Ma, E. Effects of nanocrystalline and ultrafine grain sizes on constitutive behavior and shear bands in iron. *Acta Mater.* **2003**, *51*, 3495–3509. [CrossRef]
28. Bolaños, J.A.M.; Cobos, O.F.H.; Marrero, J.M.C. Strain hardening behavior of armco iron processed by ECAP. *IOP Conf. Ser. Mater. Sci. Eng.* **2014**, *63*, 012143. [CrossRef]
29. Ding, Y.; Jiang, J.; Shan, A. Plastic instability and strain rate sensitivity of ultrafine-grained iron. *J. Alloys Compd.* **2009**, *487*, 517–521. [CrossRef]
30. Takaki, S.; Kawasaki, K.; Futamura, Y.; Tsuchiyama, T. Deformation behavior of ultrafine grained iron. *Mater. Sci. Forum* **2006**, *503–504*, 317–322. [CrossRef]
31. Legros, M.; Elliott, B.R.; Rittner, M.N.; Weertman, J.R.; Hemker, K.J. Microsample tensile testing of nanocrystalline metals. *Philos. Mag. A* **2000**, *80*, 1017–1026. [CrossRef]
32. Ma, E. Instabilities and ductility of nanocrystalline and ultrafine-grained metals. *Scr. Mater.* **2003**, *49*, 663–668. [CrossRef]
33. Cai, B.; Ma, X.; Moering, J.; Zhou, H.; Yang, X.; Zhu, X. Enhanced mechanical properties in Cu–Zn alloys with a gradient structure by surface mechanical attrition treatment at cryogenic temperature. *Mater. Sci. Eng. A* **2015**, *626*, 144–149. [CrossRef]
34. Cheng, S.; Milligan, W.W.; Wang, X.L.; Choo, H.; Liaw, P.K. Compressive and tensile deformation behavior of consolidated Fe. *Mater. Sci. Eng. A* **2008**, *493*, 226–231. [CrossRef]
35. Wang, Y.M.; Wang, K.; Pan, D.; Lu, K.; Hemker, K.J.; Ma, E. Microsample tensile testing of nanocrystalline copper. *Scr. Mater.* **2003**, *48*, 1581–1586. [CrossRef]
36. Nieman, G.W.; Weertman, J.R.; Siegel, R.W. Mechanical behavior of nanocrystalline metals. *Nanostructured Mater.* **1992**, *1*, 185–190. [CrossRef]
37. Gertsman, V.Y.; Valiev, R.Z.; Akhmadeev, N.A.; Mishin, O.V. Deformation behavior of ultrafine-grained materials. *Mater. Sci. Forum* **1996**, *225–227*, 739–744. [CrossRef]
38. Sanders, P.G.; Eastman, J.A.; Weertman, J.R. Elastic and tensile behavior of nanocrystalline copper and palladium. *Acta Mater.* **1997**, *45*, 4019–4025. [CrossRef]
39. Lu, L.; Wang, L.B.; Ding, B.Z.; Lu, K. High-tensile ductility in nanocrystalline copper. *J. Mater. Res.* **2000**, *15*, 270–273. [CrossRef]
40. Valiev, R.Z.; Kozlov, E.V.; Ivanov, Y.F.; Lian, J.; Nazarov, A.A.; Baudelet, B. Deformation behaviour of ultra-fine-grained copper. *Acta Metall. Mater.* **1994**, *42*, 2467–2475. [CrossRef]

Chapter 4:
UFG Aluminium Alloys

![metals logo] *metals*

MDPI

Article

Enhanced Mechanical Properties and Electrical Conductivity in Ultrafine-Grained Al 6101 Alloy Processed via ECAP-Conform

Maxim Murashkin [1,*], Andrey Medvedev [1], Vil Kazykhanov [1], Alexander Krokhin [2], Georgy Raab [1], Nariman Enikeev [3] and Ruslan Z. Valiev [1]

[1]   Institute for Physics of Advanced Materials, Ufa State Aviation Technical University, Ufa 450000, Russia; deckard@list.ru (A.M.); vk@mail.rb.ru (V.K.); giraab@mail.ru (G.R.); rzvaliev@mail.rb.ru (R.Z.V.)
[2]   United Company RUSAL, Moscow 109240, Russia; aleksandr.krokhin@rusal.com
[3]   Laboratory for Mechanics of Bulk Nanostructured Materials, Saint Petersburg State University, Saint-Petersburg 199034, Russia; carabus@mail.rb.ru
*   Author to whom correspondence should be addressed; m.murashkin.70@gmail.com; Tel.: +7-3472-734449; Fax: +7-3472-733422.

Academic Editor: Heinz Werner Höppel
Received: 17 September 2015; Accepted: 17 November 2015; Published: 20 November 2015

Abstract: This paper studies the effect of equal channel angular pressing-Conform (ECAP-C) and further artificial aging (AA) on microstructure, mechanical, and electrical properties of Al 6101 alloy. As is shown, ECAP-C at 130 °C with six cycles resulted in the formation of an ultrafine-grained (UFG) structure with a grain size of 400–600 nm containing nanoscale spherical metastable β' and stable β second-phase precipitates. As a result, processed wire rods demonstrated the ultimate tensile strength (UTS) of 308 MPa and electrical conductivity of 53.1% IACS. Electrical conductivity can be increased without any notable degradation in mechanical strength of the UFG alloy by further AA at 170 °C and considerably enhanced by additional decomposition of solid solution accompanied by the formation of rod-shaped metastable β' precipitates mainly in the ultrafine grain interior and by the decrease of the alloying element content in the Al matrix. It is demonstrated that ECAP-C can be used to process Al-Mg-Si wire rods with the specified UFG microstructure. The mechanical strength and electrical conductivity in this case are shown to be much higher than those in the industrial semi-finished products made of similar material processed by the conventional T6 or T81 treatment.

Keywords: Al alloy; severe plastic deformation; equal channel angular pressing-Conform; ultrafine grained structure; aging; strength; electrical conductivity

1. Introduction

Mechanical strength and electrical conductivity are the most important properties of conducting materials. Al-Mg-Si alloys, such as 6101, 6201, *etc.*, demonstrate an enhanced combination of these properties, and are currently widely used to produce electrical and power conductors of self-supporting insulated lines and overhead power transmission lines [1–5]. The materials are also considered for application as electrical wiring in the car industry [6].

Due to high corrosion resistance, specific strength, and good machinability, as well as lower cost as compared to copper and copper-based alloy conductors, the Al-Mg-Si alloys have become increasingly popular. Thus, the effort to enhance the performance properties of the materials is of particular interest to the scientific and engineering communities in terms of the use of such materials for advanced applications.

As is known, the current-conducting elements made of commercial Al-Mg-Si alloys in the form of a wire of various diameters have yield strength of 275–330 MPa and electrical conductivity of

57.5%–52.0% IACS (International Annealed Copper Standard), correspondingly [2,5]. These properties are achieved through conventional thermo-mechanical processing T81, including quenching, cold drawing, and artificial aging in a sequential order [2,7,8]. The processed microstructure has mainly non-recrystallized grains elongated in the direction of drawing. These grains contain an increased dislocation density and Mg_2Si nanoscale metastable β'' and/or β' second-phase strengthening precipitates formed along three equivalent directions <001> in the Al matrix of several nm in diameter and up to 100 nm in length [9–14]. The specified level of the material strength can be achieved through dislocation strengthening and precipitation hardening, with enhanced electrical conductivity attained via the decrease in the content of solute Mg and Si atoms in the Al matrix.

As a rule, metallurgical techniques [8,15–17] or modifications of conventional thermo-mechanical processing, for example T6 or T81 [7,18], are currently used to improve the properties of commercial Al-Mg-Si alloys. In [8], it is recommended to modify the cast Al 6201 alloy with boron microadditives to enhance electrical conductivity by decreasing the content of such impurities as Ti, Cr, V, and Mn in an Al solid solution. In [15], the possibility to increase the operating temperature of Al-Mg-Si conductors via additional alloying with Zr and the formation of Al_3Zr dispersed particles is demonstrated. In [16,17], it was shown that the increase in Si excess content could enhance the effect of precipitation hardening due to the formation of nanoscale Si precipitates along with basic strengthening Mg_2Si phases during artificial ageing. However, these techniques cannot enhance both electrical conductivity and strength of conductor materials simultaneously.

Recently, we have proposed a new approach to enhance the properties of commercial Al alloys of the Al-Mg-Si system allowing us to increase both strength and electrical conductivity of the materials. It can be achieved through combination of the UFG structure formation and solid solution decomposition accompanied by the formation of strengthening Mg_2Si phase precipitates of both metastable and stable modifications using severe plastic deformation (SPD) [19–23]. The strengthening up to 365 MPa can be achieved via grain refinement during SPD, as well as through precipitation hardening induced by following dynamic aging. The research [19–24] has also proved that in parallel with the formation of UFG structure, SPD also considerably accelerates the decomposition of a supersaturated solid solution and contributes to a greater reduction of the content of alloying element atoms in the Al matrix of Al-Mg-Si alloys as compared to conventional processing techniques. In good agreement with the Matthiessen's rule for dilute alloys [25], this leads to the increase in electrical conductivity of UFG alloys up to more than 58% IACS [19–21]. The approach was used to process Al-Mg-Si alloys via such SPD processing techniques as high pressure torsion (HPT) and equal channel angular pressing with parallel channels (ECAP-PC). Due to certain limitations (small size of samples processed, low material utilization ratio, *etc.*) these techniques of processing are regarded as laboratory ones and cannot be applied to commercially produce UFG materials. Apart from that, the ECAP-C technique is capable of producing long-length UFG billets that could be used as ingots for industrial processes [26]. As an example, the given SPD technique allowed producing UFG structures in long-length Ti and Al alloy rods [27–30], as well as in steel ones [31], resulting in a considerable enhancement of both mechanical and service characteristics as compared to similar semi-finished billets processed via conventional techniques.

The main objective of this work is to demonstrate the potential of continuous ECAP-C technique for fabrication of high strength UFG Al 6101 alloy showing a better combination of mechanical strength and electrical conductivity compared to the commercial Al-Mg-Si alloy processed via conventional thermo-mechanical treatments, which are currently used in the electrical engineering.

2. Experimental Section

Commercial Al 6101 alloy in T1 state was used as a material for study (cooled from an elevated temperature shaping process and naturally aged) with a typical chemical content, namely: 0.58 Mg; 0.54 Si; 0.23 Fe; 0.003 Cu; 0.01 Zn; 0.012 (ΣTi + V + Cr + Mn); res. Al (wt. %). The initial material had a form of continuous cast re-draw rolled rods with a diameter of 12.5 mm. The material for investigation was supplied by UC RUSAL (Moscow, Russia).

Initial rods of 1.5 m in length were successively annealed at 550 °C, water quenched and subjected to SPD. To form the UFG structure, quenched billets were SPD-processed by ECAP-C technique. A schematic drawing of the ECAP-C machine is shown in Figure 1a. Initially, a wire rod of 12.5 mm in diameter (1) heated to the specified temperature was placed into a pressing channel comprised of a running wheel die (2), pressure arrangement working surfaces (3), and a gauge (4). Friction resistance forces a wire rod going from a running wheel die (2) into a channel formed by a pressure arrangement (3) and a gauge (4), coupling at a certain angle (ψ) with a wheel die. Shear straining occurs at an intersection of these channels (deformation zone).

ECAP-C machine was used to process wire rods under isothermal conditions at 130 °C. Intersection angle of channels (ψ) (Figure 1a) constituted 120° with six processing cycles corresponding to an equivalent strain of about 4 [32]. A wire rod was rotated around the axis by +90° after each ECAP-C cycle (route *Bc*). According to the previous studies [33,34], it is the most efficient mode to form a homogeneous UFG structure in Al alloys. ECAP-C treatment resulted in the production of rods up to 1.5 m in length with a square cross section of 11 × 11 mm (Figure 1b). No noticeable macro-defects resulting from a deformation processing were observed on the surface (Figure 1b). Some part of the processed wire rods was artificially aged (AA) at the temperatures of 170 °C for 1–6 h and at 190 °C for 1–12 h.

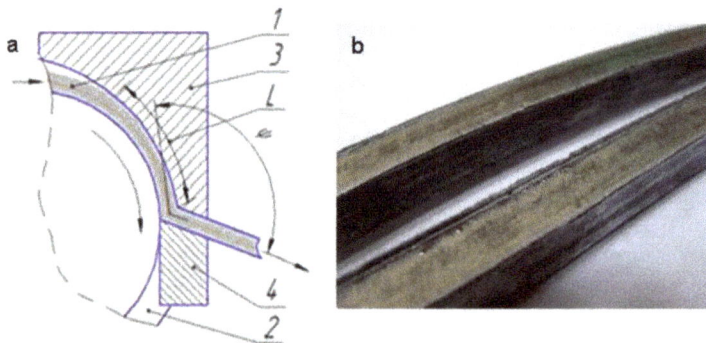

Figure 1. (**a**) A schematic illustration of ECAP-C machine: 1—ingot; 2—running wheel die; 3—pressure arrangement working surface; 4—a gauge; ψ—angle of intersection with a wheel die; *L*—arc of a rod grip; and (**b**) the surface of wire rods of Al 6101 alloy after six cycles of ECAP-C at 130 °C.

The microstructure of wire rods was studied by transmission and scanning electron microscopy (TEM and SEM). A longitudinal section of wire rods in both the initial state and processed by ECAP-C and AA was analyzed.

TEM investigations were implemented using JEM-2100 electron microscope (JEOL, Tokyo, Japan) at an accelerating voltage of 200 kV. To study the microstructure, thin foils were used, produced by jet polishing on a Tenupol-5 machine (Struers, Ballerupcity, Denmark) with the chemical solution consisting of 20% nitric acid and 80% methanol at a temperature of −25 °C and a voltage of 15 V. A mean size of structural elements was determined based on the measurements of at least 200 mean diameters. At least three foils of each state were studied to obtain statistically significant results.

The microstructure homogeneity as well as the distribution of particles of crystallized secondary phases were estimated by SEM using JSM-6490LV (JEOL, Tokyo, Japan) at an accelerating voltage of 20 kV. The chemical composition of secondary phases was analyzed by EDX technique using an extension to INCA electron microscope X-Act by the Oxford Instruments Company, Oxford, UK.

A longitudinal section of the rod in the initial state, as well as processed by SPD and further AA was analyzed.

X-ray studies were performed using Ultima IV diffractometer (Rigaku, Tokyo, Japan) by CuKα irradiation (30 kV and 20 mA). The size of coherent domains (D), values of mean-square microdistortion of crystalline lattice ($<\varepsilon^2>^{1/2}$) and crystalline lattice parameter (a) were calculated using Rietveld analysis with the help of MAUD software (copyright by L. Lutterotti, Trento, Italy) [35]. To estimate dislocation density (ρ), the Equation (1) was used [36].

$$\rho = 2\sqrt{3}\left\langle\varepsilon^2\right\rangle^{1/2}/(D \times b) \tag{1}$$

where $b = a\sqrt{2}/2$ is the Burgers vector for fcc-metals, D is the coherent domain size.

Mechanical testing was performed at room temperature using Instron 5982 tensile testing machine (Instron Engineering Corporation, Buckinghamshire, UK) with a strain rate of 10^{-3} s^{-1}. Strength (yield strength (YS) and ultimate tensile strength (UTS)) and ductility characteristics (elongation to failure (El.)) of rods were estimated based on tensile testing of the samples with a cylindrical gauge section having 3 mm in diameter and a length of 15 mm and wire samples—with a working part length of 250 mm. To obtain consistent results, at least three samples were tested per each data point.

Electrical resistivity of the material under study was measured in accordance with IEC 60468:1974 standard (CEN, Bruxelles, Belgium) [37]. Straightened samples of at least 1 m long in a measured part were taken.

3. Results and Discussion

3.1. Microstructural Features of the Alloy Processed via ECAP-C

TEM studies revealed an inhomogeneous microstructure with average grain size of 10 μm for Al 6101 alloy in the T1 state processed via a conventional technique (continuous casting and redraw rolling). Grains are elongated along the rolling direction (Figure 2a,b). The metal structure also contains intermetallic crystalline inclusions varying from 0.5 to 5 μm in size oriented along the rolling direction (marked by an arrow) (Figure 2a). This distribution of intermetallic particles is typical for Al semi-finished products processed via pressing or rolling [38]. EDX analysis of particles showed the presence of Fe, Si, and Al with a stoichiometric ratio of 2:2:9. This composition is typical for Al alloys of the Al-Mg-Si system containing no Mn and/or Cr [39].

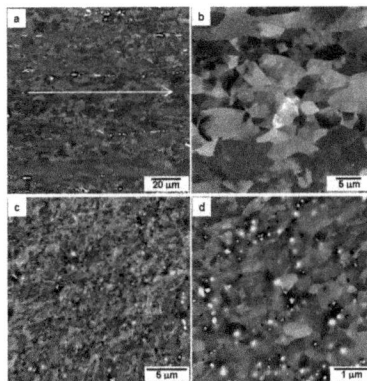

Figure 2. Backscattered electron micrographs of the microstructure of the Al 6101 alloy wire rod: (**a**, **b**) initial state (T1); and (**c**, **d**) after ECAP-C (an arrow indicated the strain direction).

A homogeneous UFG microstructure with a grain size varying from 400 to 600 nm was formed in the wire rods after six ECAP-C cycles at 130 °C (Figure 2c,d and Figure 3a,b). Grains were somewhat elongated along the shear plane. Distribution of spots along the rings on selected area electron diffraction (SAED) patterns shows the formation of grain-type structure with grain boundaries exhibiting mostly high-angle misorientation (Figure 3a) according to [33,34]. In the earlier work [40], the electron backscatter diffraction (EBSD) analysis was applied to study the evolution of grain structure during ECAP processing of an Al-Mg-Si alloy with the similar chemical composition. It was shown that ultra-fine substructure is formed after 1–2 ECAP passes and it is transformed into UFG microstructure with predominant high-angle grain boundaries after four ECAP passes and saturates with further processing.

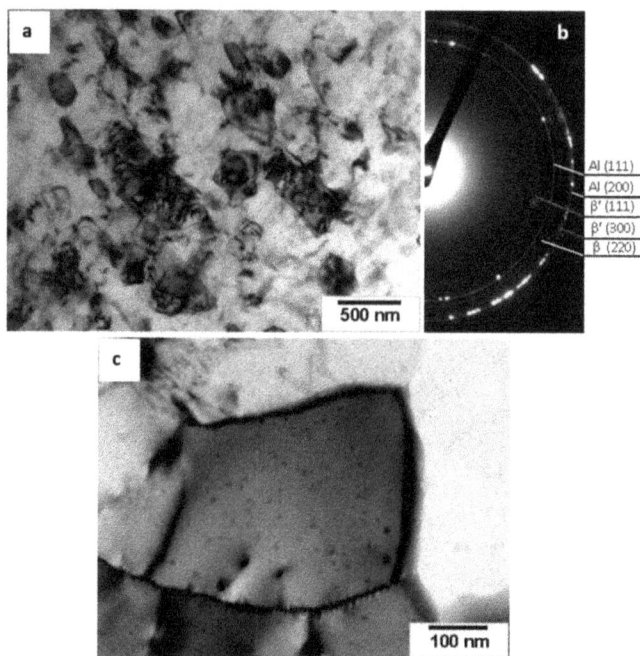

Figure 3. (**a**) Microstructure and (**b**) SAED patterns of the Al 6101 alloy after ECAP-C; and (**c**) view of strengthening second-phase precipitates formed in the matrix of the Al alloy after ECAP-C processing.

TEM studies revealed spherical disperse particles varying from 3 nm to 20 nm in size formed as a result of ECAP-C processing (Figure 3c). Their morphology same as in the previous research [19,22,41,42] on the Al-Mg-Si alloys subjected to SPD under similar temperature conditions indicates that the particles found in the Al 6101 alloy can be regarded as metastable $\beta°$-phase precipitates with hexagonal lattice with a Mg/Si ratio of 1.8 [13]. Analysis of SAED patterns (Figure 3b) shows that a part of the spots corresponding to the Al matrix, there are also spots related to the β'-precipitates, as well as β-precipitates with a Mg/Si ratio of 2. Figure 3c represents nanoscale precipitates distributed uniformly within the grain. The size of the precipitates increases considerably near GBs (Figure 3c). The same pattern of spherical β'-precipitates distribution can be observed in the UFG Al-Mg-Si alloys processed via ECAP-PC or conventional ECAP at 100 °C [22,41,42], whereas β-precipitates were observed in the Al-Mg-Si alloy processed by HPT at 130 °C [23]. These precipitates point to the fact that the UFG structure formation during ECAP-C was accompanied by solid solution decomposition due to the dynamic aging (DA). It has also been previously observed in various Al-Mg-Si alloys during SPD both at elevated [19–24,41–45] and room temperature [23,24,46].

The structure processed via ECAP-C features a high dislocation density (ρ) of about ~8×10^{13} m^{-2} according to the X-ray analysis data (Table 1). This accounts for a small size of coherent scattering regions (D), as well as for a high mean-square microdistortion of the alloy crystalline lattice ($<\varepsilon^2>^{1/2}$) (Table 1). A considerable reduction in the Al matrix lattice parameter (a) after ECAP-C is also observed as compared to rolled wire in state T1 processed via the conventional technique (Table 1).

Table 1. Results of the X-Ray diffraction analysis of the Al 6101 alloy in the initial state (T1), after ECAP-C and ECAP-C + AA.

State	D (nm)	$<\varepsilon^2>^{1/2}$ (%)	a (Å)	ρ (m^{-2})
Initial state—T1 (this work)	-	-	4.0531 ± 0.0008	-
6 cycles ECAP-C at 170 °C	165 ± 20	0.110 ± 0.024	4.0520 ± 0.0002	8.0×10^{13}
ECAP-C + AA at 170 °C	188 ± 13	0.067 ± 0.031	4.0508 ± 0.0004	4.3×10^{13}

These results are consistent with the phase composition changes observed in the wire rods after ECAP-C by TEM (Figure 3c). This points to a notable decrease in the concentration of Mg and Si in the Al matrix [19,22,42].

ECAP-C processing of a wire rod led to not only the formation of UFG structure and strengthening phase, but also to the distortion of a linear orientation of crystallized intermetallic particles (Figure 2c). A more uniform nature of their distribution was observed, which could reduce to a certain degree the anisotropy of the material mechanical properties [38]. Such a redistribution of intermetallic particles in the Al matrix is possibly the result of SPD [26].

3.2. Mechanical Properties and Electrical Conductivity of the Alloy Processed via ECAP-C

Figure 4 and Table 2 demonstrate the mechanical properties of UFG rolled wire processed via ECAP-C, as well as its properties in T1 state. The Table 2 also contains the values of specific electrical resistivity found experimentally and calculated electrical conductivity values (% IACS). Physical and mechanical properties of Al 6101 alloy in the form of rods and wire produced by different manufacturers are given in Table 2 for comparison. The results gained show that the formation of UFG microstructure in the wire rod enhances the yield strength and the yield stress by ~2.3 and 1.5 times, correspondingly, as compared to the counterparts processed via the conventional technique in state T1 and T4. The formation of UFG structure leads to not only the enhancement of mechanical strength but also to the considerable increase in electrical conductivity. Additionally, UFG wire rods are much stronger than those of the Al 6101 alloy in T6 state (Table 2).

Table 2. Mechanical properties and electrical conductivity of the Al 6101 alloy.

State	YS (MPa)	UTS (MPa)	El$_{total}$ (%)	Resistivity [3] ($\Omega \cdot$mm^2/m)	IACS (%)
Initial state—T1 (this work)	120 ± 1	195 ± 2	22.0 ± 0.4	0.03423	50.4
UFG wire rods produced by ECAP-C + AA					
6 passes ECAP-C at 130 °C	282 ± 8	308 ± 9	15.1 ± 0.6	0.03242	53.1
ECAP-C + AA at 170 °C	291 ± 10	304 ± 3	15.0 ± 0.3	0.03020	57.1
ECAP-C + AA at 190 °C	261 ± 6	288 ± 4	16.8 ± 0.9	0.03001	57.4
Wire 3.2 mm in diameter produced from UFG rods					
ECAP-C + AA at 170 °C + drawing (diam. 3.2 mm)	-	364 ± 9	3.5 ± 0.2	0.03055	56.4
Conventionally produced continuous cast redraw rolled rods					
T1 [2]	-	≥190	≥17	≤0.03500	49.3
T4 [1]	-	180–205	15	0.03316	52.0
Wire produced by conventional thermal/thermomechanical treatment					
T6 [1]	-	205–250	-	0.03135	55.0
T81 [1]	-	305–315	6	0.03253	53.0
AL2 (diam. 3.0–5.0 mm) [5]	-	315	3.0	0.03284	52.5
AL7 (diam. 3.0–3.5 mm) [5]	-	275	3.0	0.03050	56.5

T1—Cooled from an elevated temperature shaping process and naturally aged; T4—Solution heat treated and naturally aged; T6—Solution heat treated and artifically aged; T81—Solution heat treated, cold-worked and artificially aged; [1] aluminum continuous cast re-draw rod and wire produced by Southwire (Carrollton, GA, USA) [47]; [2] aluminum continuous cast re-draw rod produced by NPA Skawina (Poland) [48]; [3] electrical resistivity at 20 °C.

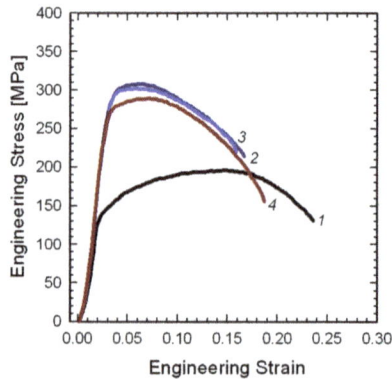

Figure 4. Typical engineering stress—engineering strain curves from tensile testing of the Al 6101 alloy (1) in the T1 condition; (2) after ECAP-C processing; (3) after ECAP-C processing and AA at 170 °C for 12 h; and (4) after ECAP-C processing and AA at 190 °C for 6 h.

3.3. Properties and Microstructure of the UFG Alloy Processed via Artificial Aging

In [22,23] it was found that a notable enhancement of electrical conductivity in UFG Al-Mg-Si alloys after SPD treatment could be achieved by further artificial aging (AA). There has also been noted that AA can lead to both strength and ductility increases in UFG alloys [22,42,44,49], or only to mechanical strength enhancement with an admissible ductility decrease [41]. Following these findings a part of the ECAP-C processed wire rod was subjected to AA at 170 °C and 190 °C—the temperatures that are typically used in traditional heat treatment procedures of the commercial rods/wire of 6101 alloy [2,6–8,50].

Figure 5 demonstrates the electrical conductivity dependence on the temperature and the time of AA for UFG wire rods. It can be seen that AA results in a considerable enhancement of electrical conductivity up to over 57% IACS (Figure 5, Table 2). It is obvious that the increase in AA temperature notably reduces the time necessary to achieve high electrical conductivity. The measurements taken revealed that the best possible combination of strength (UTS over 304 MPa), ductility (15.1%), and electrical conductivity (57.1% IACS) are demonstrated by the UFG wire rods processed by AA at 170 °C for 12 h. The given level of properties is considerably higher than that of the alloy processed via the conventional techniques (Table 2).

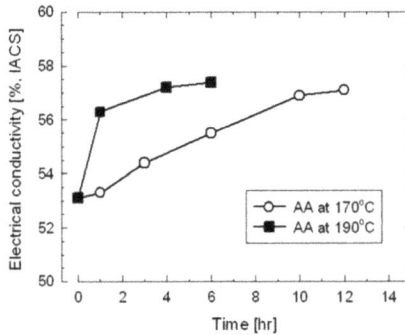

Figure 5. Change in electrical conductivity of test samples of the Al 6101 rod wire after ECAP-C and AA over time.

TEM microstructural studies of the wire rods after ECAP-C and further AA revealed the same grain size as for the SPD-processed material. Metastable β′-second-phase precipitates formed in the grain interior following aging are of a spherical form with a diameter of up to 5 nm and with a length varying from 10 to 50 nm oriented along <001> direction of the Al matrix (Figure 6). A corresponding typical SAED pattern is presented in Figure 6b. Symmetrical streaks between the diffraction spots of [001] Al are related to HCP β′-precipitates [51,52]. Their morphology and location are typical of coarse-grained alloys of the Al-Mg-Si system processed by quenching and further AA [9–13]. Precipitates of the same morphology were observed after AA in UFG Al-Mg-Si alloys processed via ECAP-PC [19,22,42] and HPT [23]. In the research [19,22,42], it was revealed that the grain interior contained not only β′-precipitates, but also needle-shaped metastable β″-phase. The absence of this phase in UFG 6101 alloy in the current study may result from the application of higher temperature range of AA. The change in the UFG alloy phase composition is accompanied by a considerable (almost by two times) decrease in $<\varepsilon^2>^{1/2}$ and ρ values (Table 1). Such variations with no notable change in a grain size suggest that recovery processes take place in the UFG structure [26,34]. Additionally, a considerable decrease in the crystalline lattice parameter (a) of the Al matrix is noted (Table 1). The parameter comes close to the value for pure Al (4.0495 A). The found variation of a points to the further decomposition of solid solution accompanied by the growth of precipitations already formed during ECAP-C and the formation of the new ones. The observed decrease in $<\varepsilon^2>^{1/2}$, ρ, and, particularly, in a, is the reason for high electrical conductivity of UFG wire rods (Table 2). Such an interrelation between a parameter and electrical conductivity has already been established in the studies [19,21–23].

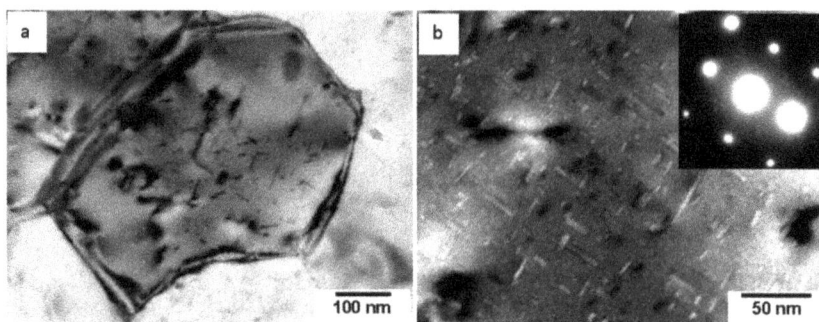

Figure 6. Strengthening second-phase precipitates present in the Al matrix of the ECAP-C processed alloy artificially aged at 170 °C for 12 h: (**a**) bright field TEM image; and (**b**) bright field TEM image and corresponding SAED patterns (the crystal is close to the [001] zone axis orientation).

At present, SPD techniques aimed at processing of commercially applicable UFG metals and alloys with enhanced mechanical and functional characteristics are actively developed [26,53]. The given research demonstrates the possibility to process the specified UFG structure having a unique combination of mechanical strength and electrical conductivity in rods made of Al 6101 alloy using the ECAP-C technique with further artificial aging.

The first processing stage included six cycles (equivalent strain of ~4) of ECAP-C at 130 °C by Bc route after quenching. The selection of the SPD regime is determined by several reasons.

Firstly, the previous studies showed that SPD processing of Al-Mg-Si alloys under similar conditions led to the formation of a homogeneous UFG structure with enhanced mechanical properties [21,22].

Secondly, it was found that elevated temperature of SPD in Al alloys resulted in a considerable decomposition of solid solution during dynamic aging (DA) accompanied by the formation of nanoscale second-phase precipitates. This led to additional enhancement of both strength and electrical conductivity [21,23].

Thirdly, the selected regimes of SPD provided the capacity for further aging to enhance the properties of UFG alloys. It was shown in the following studies [22,23].

Fourthly, the selected conditions of SPD assure the fabrication of long-length rods with no noticeable macro and/or micro defects.

ECAP-C of wire rods of up to 1.5 m in length resulted in the formation of a homogeneous UFG structure from 400 to 600 nm in size. Additionally, nanoscale spherical metastable β'-second-phase precipitates were formed in the Al matrix during ECAP-C. Their morphological features are typical for Al alloys, since the formation of ultrafine grains in these alloys is accompanied by the decomposition of solid solution and the formation of strengthening Mg_2Si phases [19–24,41–46]. Their spherical shape is different from the shape of β'-phase (in the form of spheres) formed as a result of a conventional thermal (of T6 type) and/or thermomechanical treatment (of T81 type) of CG counterparts [9–14]. The observed differences are determined by a considerable dislocation activity and the increase in vacancy concentration during SPD (ECAP-C in our case). This leads to a notable increase in the effective diffusion factor in Al alloys [10,54]. The diffusion stepping up accelerates the precipitation kinetics and contributes to the formation of metastable β'-phase after ECAP-C at 130 °C already.

Judging by the character of metastable phase distribution shown in Figure 3b it can be assumed that grain-boundary diffusion plays an important role in the formation and growth of β'-phase metastable precipitates found mainly along GBs. It is well known that the grain-boundary diffusion coefficient is several orders greater than the volume diffusion coefficient [55]. Consequently, the kinetics of the α_{Al}-GB-β''-β' phase transformation and further enlargement of β'-phase particles formed along

GBs is suggested to be controlled by grain-boundary diffusion and is much more enhanced than similar processes inside grains controlled by volume diffusion. Similar character of metastable phase precipitates distribution in UFG Al alloys of the Al-Mg-Si system after SPD treatment was reported in [19,22,23,42].

At the second processing stage of UFG wire rods (after ECAP-C), further AA resulted in the formation of β'-phase precipitates mainly inside grains (Figure 6). This is due to the fact that solute atoms are mostly concentrated inside grains as compared to the regions adjacent to GBs. A similar distribution of metastable phase precipitates was observed in a number of UFG Al-Mg-Si alloys after AA [19,21–23,42]. It is necessary to note that both the current study and the previous research point to the fact that the morphology of metastable phase precipitates in UFG alloys is similar to that of CG counterparts processed by conventional techniques [9–14]. The absence of metastable β''-phase precipitates in the UFG alloy after AA found earlier in Al-Mg-Si alloys after SPD treatment [19,21,22,42] can be explained by the higher aging temperature used to process billets.

The study of the mechanical properties of the UFG alloy shows that the results obtained (Table 2) are in good agreement with microstructural changes taking place after both ECAP-C and further AA. The formation of homogeneous UFG structure (Figures 3 and 6) with mainly large-angle grain boundary misorientations in Al 6101 alloy wire rods after six cycles of ECAP-C at 130 °C resulted in an expected strength enhancement as defined by the Hall-Petch relationship [56,57]. It has been repeatedly confirmed for Al alloys in the UFG state [26,34]. The formation of spherical metastable β'-phase precipitates of ultrafine grains (Figure 4) during AA at 170 and 190 °C does not have any effect on the mechanical properties and even degrades them a little (Table 2). With no noticeable grain enlargement following the aging, this modification of properties in the UFG materials is connected with two competing processes. These are (i) the recovery caused by a considerable decrease in dislocation density (by almost two times) and (ii) the precipitation hardening, due to the formation of metastable β'-phase precipitates inside grains in the course of AA. It was also noted than no strengthening or even certain strength degradation took place in UFG Al-Mg-Si alloys as a result of AA [42]. Spherical β'-phase precipitates formed along GBs of the UFG structure as a result of dynamic aging (DA) play no part in total strengthening of the material [58].

It is well known that electrical conductivity depends on microstructural features of metallic materials determined by electron scattering as a consequence of crystalline structure distortions, including atom thermal vibrations, particles, crystalline lattice defects, *etc.* Thus, it seems very complicated to enhance strength of conductor materials without a significant deterioration of their electrical conductivity using conventional techniques [25]. For example, alloying of pure metals, their strain or precipitation hardening results in a considerable decrease of electrical conductivity due to a notable expansion of electron scattering on grain/sub-grain boundaries, increased amount of solute atoms in the matrix, GP zones, secondary precipitates, and dislocations. It is known that solute atoms in the matrix and GP zones are more efficient in scattering of electrons than other structural elements [25,56]. SPD-processing leads to grain refinement followed by solid solution decomposition and subsequent precipitation and can be used as a new approach to enhance mechanical strength of pure metals with retention of electrical conductivity. It was applied to pure Cu [59,60], and recently has been used to process Cu [61,62] and Al-based alloys [19–23,63]. In our case, DA, in the course of ECAP-C, results in solid solution decomposition, which contributes to certain electrical conductivity enhancement (Table 2). The same effect of DA during SPD was observed in our previous studies [19–23,61]. The dependence of electrical conductivity of UFG Al-Mg-Si alloys on stable Mg_2Si phase precipitates and precipitates of its metastable modifications [19–23] was analyzed. The obtained results suggest that secondary β' precipitates formed both inside grains and along GBs following ECAP-C have insignificant negative effect on the electrical conductivity of the material under study, while providing significant strengthening effect. Higher amount of Mg atoms are required for formation of β' precipitates (Mg/Si ratio of 1.8) [13] in comparison with β'' precipitates (Mg/Si ratio of 1.1) [64]. Therefore, β' precipitates effectively purify the Al matrix from Mg solute atoms, which most negatively affect the electrical conductivity of the material [65]. The formation of new secondary precipitates in the course of AA leads to a further matrix depletion with alloying elements and a more considerable enhancement of UFG 6101 alloy electrical conductivity (Figure 6, Table 2).

Thus, the UFG Al-Mg-Si processed via six ECAP-C cycles at 130 °C and AA at 170 or 190 °C has electrical conductivity essentially higher than that of counterparts subjected to conventional treatment techniques (Table 2).

It is necessary to note that the unique combination of high mechanical strength and electrical conductivity was achieved in long-length rods through application of ECAP-C. This method appears to be particular promising, especially taking into account first pilot wire samples (Table 2) with an enhanced combination of properties produced from UFG rods by cold drawing modes typically used commercially. Pilot UFG wire samples have UTS of over 360 MPa, elongation to failure of over 3% and electrical conductivity of 56.4% IACS. These studies demonstrate the practical feasibility of a fast transition from pilot rods and wire to their commercial production [66].

4. Conclusions

1. It is demonstrated for the first time that continuous ECAP-C processing can be efficiently used to form the UFG microstructure in Al alloys of the Al-Mg-Si system. ECAP-C with six cycles at 130 °C of Al 6101 alloy resulted in the formation of a homogeneous UFG microstructure with a grain size varying from 400 nm to 600 nm in long-length rods.

2. Grain refinement down to ultrafine scale during ECAP-C is accompanied by a dynamic aging leading to the formation of spherical metastable β'-second-phase precipitates and stable β-phase from 3 nm to 20 nm in size.

3. Long-length wire rods made of UFG Al 6101 alloy have much higher mechanical strength as compared to their counterparts processed via conventional treatment techniques. Further artificial aging at 170 °C of UFG billets processed via ECAP-C has no effect on their mechanical strength (at 170 °C) but leads to some strength degradation at 190 °C.

4. Electrical conductivity of the Al 6101 alloy depends on its microstructural features due to SPD processing. Decomposition of supersaturated solid solution leading to the formation of metastable β'-phase and stable β-phase in the course of ECAP-C as a result of dynamic aging and additional β'-phase during artificial aging is an efficient approach to enhance the electrical conductivity of the alloy. UFG long-length rods were used to produce the wire samples by cold drawing with superior properties considerably exceeding the existing counterparts processed via conventional techniques.

Acknowledgments: Authors would like to acknowledge gratefully the Russian Ministry for Education and Science for the financial support of this study through the Federal Targeted Program "The development of a technology for producing nanostructured Al alloys with increased constructional strength for engineering electrical application", ID 2014-14-579-0069-003.

Author Contributions: Maxim Murashkin formulated the idea of this work, supervised the workflow and created the initial draft. Andrei Medvedev carried out mechanical tensile tests, conductivity measurements and performed microstructural characterization of the processed material by SEM. Vil Kazykhanov performed microstructural characterization studies of the processed material by TEM, Nariman Enikeev performed microstructural characterization with the help of the X-ray diffraction analysis. Alexandr Krokhin supplied the initial material for study, Georgi Raab processed the wire rods by ECAP-C. Ruslan Z. Valiev contributed with the overall supervision and development of the main concepts presented in this paper. All authors discussed the experimental results, participated in manuscript preparation and approved the final manuscript.

Conflicts of Interest: The authors declare no conflict of interest.

References

1. Nicoud, J.C.; Gaschingard, P. Important recent developments in the manufacture of redraw rod in Almelec (AA-6101 Al-Mg-Si Alloy) for the conductors employed on overhead power transmission and distribution lines. In Proceedings of the Wire Assoc, Asia 82, Singapore, 4–8 October 1982; pp. 43–46.

2. Davies, G. Aluminium alloy (6201, 6101A) conductors. In Proceedings of the International Conference on Overhead Line Design and Construction: Theory and Practice (up to 150 kV), London, UK, 28–30 November 1988; pp. 93–97.

3. Kiessling, F.; Nefzger, P.; Nolasco, J.F.; Kaintzyk, U. *Overhead Power Lines: Planning, Desing, Construction*; Springer-Verlag: Berlin, Germany, 2003.

4. Totten, G.E.; MacKenzie, D.S. (Eds.) *Handbook of Aluminium*; Marcel Dekker: New York, NY, USA, 2003.

5. European Committee for Standardization (CEN). *EN 50183, Overhead Power Line Conductors—Bare Conductors of Aluminium Alloy with Magnesium and Silicon Content*; CEN: Bruxelles, Belgium, 2002.

6. Koch, S. Aluminum alloys for wire harnesses in automotive engineering. *BHM* **2007**, *152*, 62–67. [CrossRef]

7. Cervantes, E.; Guerrero, M.; Ramos, J.A.; Montes, S.A. Influence of natural aging and cold deformation on the mechanical and electrical properties of 6201-T81 aluminum alloy wires. *Mater. Res. Soc. Symp. Proc.* **2010**, *1275*, 75–80. [CrossRef]

8. Karabay, S. Modification of AA-6201 alloy for manufacturing conductivity and extra high conductivity wires with property of high tensile stress after artificial aging heat treatment of high for all aluminium alloy conductors. *Mater. Design* **2006**, *27*, 821–832. [CrossRef]

9. Edwards, G.A.; Stiller, K.; Dunlop, G.L.; Couper, M.J. The precipitation sequence in Al-Mg-Si alloys. *Acta Mater.* **1998**, *46*, 3893–3904. [CrossRef]

10. Murayama, M.; Hono, K.; Saga, M.; Kikuchi, M. Atom probe studies on the early stages of precipitation in Al-Mg-Si alloys. *Mater. Sci. Eng. A* **1998**, *250*, 127–132. [CrossRef]

11. Murayama, M.; Hono, K. Pre-precipitate clusters and precipitation processes in Al-Mg-Si alloys. *Acta Mater.* **1999**, *47*, 1537–1548.

12. Andersen, S.J.; Marioara, C.D.; Froseth, A.; Vissers, R.; Zandbergen, H.W. Crystal structure of the orthorhombic U2-Al$_4$Mg$_4$Si$_4$ precipitate in the Al-Mg-Si alloy system and its relation to the β' and β'' phases. *Mater. Sci. Eng. A* **2005**, *390*, 127–138. [CrossRef]

13. Vissers, R.; van Huis, M.A.; Jansen, J.; Zandbergen, H.W.; Marioara, C.D.; Andersen, S.J. The Crystal Structure of the β' Phase in Al-Mg-Si Alloys. *Acta Mater.* **2007**, *55*, 3815–3823. [CrossRef]

14. Pogatscher, S.; Antrekowitsch, H.; Leitner, H.; Ebner, T.; Uggowitzer, P.J. Mechanisms controlling the artificial aging of Al-Mg-Si Alloys. *Acta Mater.* **2011**, *59*, 3352–3363. [CrossRef]

15. Wuhua, Y.; Zhenyu, L. Effect of Zr addition on properties of Al-Mg-Si aluminum alloy used for all aluminum alloy conductor. *Mater. Design* **2011**, *32*, 4195–4200.

16. Gupta, A.K.; Lloyd, D.J.; Court, S.A. Precipitation hardening processes in an Al-0.4% Mg-1.3% Si-0.25% Fe aluminum alloy. *Mater. Sci. Eng. A* **2001**, *301*, 140–146. [CrossRef]

17. Gupta, A.K.; Lloyd, D.J.; Court, S.A. Precipitation hardening in Al-Mg-Si alloys with and without excess Si. *Mater. Sci. Eng. A* **2001**, *316*, 11–17. [CrossRef]

18. Liu, C.H.; Chen, J.; Lai, Y.X.; Zhu, D.H.; Gu, Y.; Chen, J.H. Enhancing electrical conductivity and strength in Al alloys by modification of conventional thermo-mechanical process. *Mater. Design* **2015**. [CrossRef]

19. Bobruk, E.V.; Murashkin, M.Y.; Kazykhanov, V.U.; Valiev, R.Z. Aging behavior and properties of ultrafine-grained aluminum alloys of Al-Mg-Si system. *Rev. Adv. Mater. Sci.* **2012**, *31*, 101–115.

20. Sabirov, I.; Murashkin, M.; Valiev, R.Z. Nanostructured aluminium alloys produced by severe plastic deformation: New horizons in development. *Mater. Sci. Eng. A* **2013**, *560*, 1–24. [CrossRef]

21. Valiev, R.Z.; Murashkin, M.Y.; Sabirov, I. A nanostructural design to produce high-strength Al alloys with enhanced electrical conductivity. *Scr. Mater.* **2014**, *76*, 13–16. [CrossRef]

22. Murashkin, M.; Sabirov, I.; Kazykhanov, V.; Bobruk, E.; Dubravina, A.; Valiev, R.Z. Enhanced mechanical properties and electrical conductivity in ultra-fine grained Al alloy processed via ECAP-PC. *J. Mater. Sci.* **2013**, *48*, 4501–4509. [CrossRef]

23. Sauvage, X.; Bobruk, E.V.; Murashkin, M.Y.; Nasedkina, Y.; Enikeev, N.A.; Valiev, R.Z. Optimization of electrical conductivity and strength combination by structure design at the nanoscale in Al-Mg-Si alloys. *Acta Mater.* **2015**, *98*, 355–366. [CrossRef]

24. Sha, G.; Tugcu, K.; Liao, X.Z.; Trimby, P.W.; Murashkin, M.Y.; Valiev, R.Z.; Ringer, S.P. Strength, grain refinement and solute nanostructures of an Al-Mg-Si alloy (AA6060) processed by high-pressure torsion. *Acta Mater.* **2014**, *63*, 169–179. [CrossRef]

25. Rositter, P.L. *The Electrical Resistivity of Metals and Alloys*; Cambridge University Press: Cambridge, UK, 2003.

26. Valiev, R.Z.; Zhilyaev, A.P.; Langdon, T.G. *Bulk Nanostructured Materials: Fundamentals and Applications*; John Wiley & Sons, Inc.: Hoboken, NJ, USA, 2014.

27. Semenova, I.P.; Polyakov, A.V.; Raab, G.I.; Lowe, T.C.; Valiev, R.Z. Enhanced fatigue properties of ultrafine-grained Ti rods processed by ECAP-Conform November. *J. Mater. Sci.* **2012**, *47*, 7777–7781. [CrossRef]

28. Gunderov, D.V.; Polyakov, A.V.; Semenova, I.P.; Raab, G.I.; Churakova, A.A.; Gimaltdinova, E.I.; Sabirov, I.; Segurado, J.; Sitdikov, V.D.; Alexandrov, I.V.; *et al.* Evolution of microstructure, macrotexture and mechanical properties of commercially pure Ti during ECAP-conform processing and drawing. *Mater. Sci. Eng. A* **2013**, *562*, 128–136. [CrossRef]

29. Raab, G.J.; Valiev, R.Z.; Lowe, T.C.; Zhu, T.Y. Continuous processing of ultrafine grained Al by ECAP-Conform. *Mater. Sci. Eng. A* **2004**, *382*, 30–34. [CrossRef]

30. Xu, C.; Schroeder, S.; Berbon, P.B.; Langdon, T.G. Principles of ECAP-Conform as a continuous process for achieving grain refinement: Application to an aluminum alloy. *Acta Mater.* **2010**, *58*, 1379–1386. [CrossRef]

31. Niendorf, T.; Böhner, A.; Höppel, H.W.; Göken, M.; Valiev, R.Z.; Maier, H.J. Comparison of the monotonic and cyclic mechanical properties of ultrafine-grained low carbon steels processed by continuous and conventional equal channel angular pressing. *Mater. Design* **2013**, *47*, 138–142. [CrossRef]

32. Segal, V.M. Materials processing by simple shear. *Mater. Sci. Eng. A* **1995**, *147*, 157–164. [CrossRef]

33. Langdon, T.G.; Furukawa, M.; Nemoto, M.; Horita, Z. Using equal-channel angular pressing for refining grain size. *JOM* **2000**, *52*, 30–33. [CrossRef]

34. Valiev, R.Z.; Langdon, T.G. Principles of equal-channel angular pressing as a processing tool for grain refinement. *Prog. Mater. Sci.* **2006**, *51*, 881–981. [CrossRef]

35. Lutterotti, M.; Matthies, S.; Wenk, H.R. MAUD (Material Analysis Using Diffraction): A user friendly Java program for Rietveld Texture Analysis and more. In Proceeding of the 12th International Conference on Textures of Materials (ICOTOM-12), Montreal, QC, Canada, 9–13 August 1999; p. 1599.

36. Williamson, L.K.; Smallman, R.E. III. Dislocation densities in some annealed and cold-worked metals from measurements on the X-ray Debye–Scherrer spectrum. *Phil. Mag.* **1956**, *1*, 34–45. [CrossRef]

37. International Electrotechnical Commission (IEC). *IEC 60468: Method of Measurement of Resistivity of Metallic Materials*; IEC: Geneva, Switzerland, 1974.

38. Mondolfo, L.F. *Aluminum Alloys: Structure and Properties*; Butterworths: London, UK, 1976.

39. Hatch, J.E. *Aluminum: Properties and Physical Metallurgy*; ASM International: Metals Park, OH, USA, 1984.

40. Mackenzie, P.W.J.; Lapovok, R. ECAP with back pressure for optimum strength and ductility in aluminium alloy 6016. Part 1: Microstructure. *Acta Mater.* **2010**, *58*, 3198–3211. [CrossRef]

41. Sauvage, X.; Murashkin, M.Y.; Valiev, R.Z. Atomic scale investigation of dynamic precipitation and grain boundary segregation in a 6061 aluminium alloy nanostructured by ECAP. *Kov. Mater. Met. Mater.* **2011**, *49*, 11–15.

42. Bobruk, E.V.; Kazykhanov, V.U.; Murashkin, M.Y.; Valiev, R.Z. Enhanced strengthening in ultrafine-grained Al-Mg-Si alloys produced via ECAP with parallel channels. *AEM* **2015**. [CrossRef]

43. Roven, H.J.; Liu, M.; Werenskiold, J.C. Dynamic precipitation during severe plastic deformation of an Al-Mg-Si aluminium alloy. *Mater. Sci. Eng. A* **2008**, *483–484*, 54–58. [CrossRef]

44. Kim, W.J.; Kim, J.K.; Park, T.Y.; Hong, S.I.; Kim, D.I.; Kim, Y.S.; Lee, J.D. Enhancement of strength and superplasticity in a 6061 Al alloy processed by equal-channel-angular-pressing. *Metall. Mater. Trans. A* **2002**, *33*, 3155–3164. [CrossRef]

45. Kashyap, B.P.; Hodgson, P.D.; Estrin, Y.; Timokhina, I.; Barnett, M.R.; Sabirov, I. Plastic flow properties and microstructural evolution in an ultrafine-grained Al-Mg-Si alloy at elevated temperatures. *Metall. Mater. Trans. A* **2009**, *40*, 3294–3303. [CrossRef]

46. Nurislamova, G.; Sauvage, X.; Murashkin, M.; Islamgaliev, R.; Valiev, R. Nanostructure and related mechanical properties of an Al-Mg-Si alloy processed by severe plastic deformation. *Philos. Mag. Lett.* **2008**, *88*, 459–466. [CrossRef]

47. Southwire®. Available online: http://www.southwire.com (accessed on 25 May 2015).

48. NPA Skowina. Available online: http://www.npa.pl (accessed on 13 April 2015).

49. Roven, H.J.; Nesboe, H.; Werenskiold, J.C.; Seibert, T. Mechanical properties of aluminium alloys processed by SPD: Comparison of different alloy systems and possible product areas. *Mater. Sci. Eng. A* **2005**, *410–411*, 426–429. [CrossRef]

50. Aluminum Association. *Aluminum Standarts and Date*, 2nd ed.; Aluminum Association: Washington, DC, USA, 1986.

51. Dutta, I.; Allen, S.M. A calorimetric study of precipitation in commercial aluminium alloy 6061. *J. Mater. Sci. Lett.* **1991**, *10*, 323–326. [CrossRef]

52. Cai, M.; Field, D.P.; Lorimer, G.W. A systematic comparison of static and dynamic ageing of two Al-Mg-Si alloys. *Mater. Sci. Eng. A* **2004**, *373*, 65–71. [CrossRef]

53. Fakhretdinova, E.; Raab, G.; Ryzhikov, O.; Valiev, R. Processing ultrafine-grained aluminum alloy using Multi-ECAP-Conform technique. *IOP Conf. Ser. Mater. Sci. Eng.* **2014**. [CrossRef]

54. Genevois, C.; Fabregue, D.; Deschamps, A.; Poole, W.J. On the coupling between precipitation and plastic deformation in relation with friction stir welding of AA2024 T3 aluminium alloy. *Mater. Sci. Eng. A* **2006**, *441*, 39–48. [CrossRef]

55. Frost, H.J.; Ashby, M.F. *Deformation-Mechanism Maps: The Plasticity and Creep of Metals and Ceramics*; Pergamon Press: Oxford, UK, 1982.

56. Hall, E.O. The deformation and ageing of mild steel: III Discussion of results. *Proc. Phys. Soc.* **1951**, *64*, 747–752. [CrossRef]

57. Petch, N.J. The cleavage strength of polycrystals. *J. Iron Steel Inst.* **1953**, *174*, 25–28.

58. De, P.S.; Su, J.Q.; Mishra, R.S. A stress-strain model for a two-phase ultrafine-grained aluminum alloy. *Scr. Mater.* **2011**, *64*, 57–60. [CrossRef]

59. Takata, N.; Lee, S.H.; Tsuji, N. Ultrafine grained copper alloy sheets having both high strength and high electric conductivity. *Mater. Lett.* **2009**, *63*, 1757–1760. [CrossRef]

60. Zhang, Y.; Li, Y.S.; Tao, N.R.; Lu, K. High strength and high electrical conductivity in bulk nano-grained Cu embedded with nano-scale twins. *App. Phys. Lett.* **2007**. [CrossRef]

61. Islamgaliev, R.K.; Nesterov, K.M.; Bourgon, J.; Champion, Y.; Valiev, R.Z. Nanostructured Cu-Cr alloy with high strength and electrical conductivity. *J. Appl. Phys.* **2014**. [CrossRef]

62. Valiev, R.Z.; Sabirov, I.; Zhilyaev, A.P.; Langdon, T.G. Bulk nanostructured metals for innovative applications. *JOM* **2012**, *64*, 1134–1142. [CrossRef]

63. Murashkin, M.Y.; Sabirov, I.; Sauvage, X.; Valiev, R.Z. Nanostructured Al and Cu alloys with superior strength and electrical conductivity. *J. Mater. Sci.* **2015**. [CrossRef]

64. Hasting, H.S.; Frøseth, A.G.; Andersen, S.J.; Vissers, R.; Walmsley, J.C.; Marioara, C.D.; Danoix, F.; Lefebvre, W.; Holmestad, R. Composition of β″ precipitates in Al-Mg-Si alloys by atom probe tomography and first principles calculations. *J. Appl. Phys.* **2009**. [CrossRef]

65. Kutner, F. *Aluminium—Monograph, Aluminium Conductor Materials*; Aluminium-Verlag GmbH: Duesseldorf, Germany, 1981; pp. 15–27.

66. Mann, V.K.; Krokhin, A.Y.; Matveeva, I.A.; Raab, G.I.; Murashkin, M.Y.; Valiev, R.Z. Nanostructured wire rod research and development. *Light Met. Age* **2014**, *72*, 26–29.

metals

MDPI

Article

Fatigue Behavior of an Ultrafine-Grained Al-Mg-Si Alloy Processed by High-Pressure Torsion

Maxim Murashkin [1,2,*], Ilchat Sabirov [3], Dmitriy Prosvirnin [4,†], Ilya Ovid'ko [5], Vladimir Terentiev [4,†], Ruslan Valiev [1,2] and Sergey Dobatkin [4,†]

[1] Institute for Physics of Advanced Materials, Ufa State Aviation Technical University, Ufa 450000, Russia; rzvaliev@mail.rb.ru
[2] Laboratory for Mechanics of Bulk Nanostructured Materials, Saint Petersburg State University, Saint-Petersburg 199034 Russia
[3] IMDEA Materials Institute, Getafe, Madrid 28906, Spain; ilchat.sabirov@imdea.org
[4] Baikov Institute for Metallurgy and Materials Science of the Russian Academy of Sciences, Moscow 119991, Russia; imetran@yandex.ru (D.P.); fatig@mail.ru (V.T.); dobatkin@ultra.imet.ac.ru (S.D.)
[5] Research Laboratory for Mechanics of New Nanomaterials, Saint-Petersburg State Polytechnical University, Saint-Petersburg 195251, Russia; ovidko@gmail.com
*   Author to whom correspondence should be addressed; m.murashkin.70@gmail.com; Tel.: +7-3472-734449; Fax: +7-3472-733422.
†   These authors contributed equally to this work.

Academic Editor: Heinz Werner Höppel
Received: 24 March 2015; Accepted: 31 March 2015; Published: 10 April 2015

Abstract: The paper presents the evaluation of the mechanical and fatigue properties of an ultrafine-grained (UFG) Al 6061 alloy processed by high-pressure torsion (HPT) at room temperature (RT). A comparison is made between the UFG state and the coarse-grained (CG) one subjected to the conventional aging treatment T6. It is shown that HPT processing leads to the formation of the UFG microstructure with an average grain size of 170 nm. It is found that yield strength ($\sigma_{0.2}$), ultimate tensile strength (σ_{UTS}) and the endurance limit (σ_f) in the UFG Al 6061 alloy are higher by a factor of 2.2, 1.8 and 2.0 compared to the CG counterpart subjected to the conventional aging treatment T6. Fatigue fracture surfaces are analyzed, and the fatigue behavior of the material in the high cycle and low cycle regimes is discussed.

Keywords: Al alloy; severe plastic deformation; high pressure torsion; ultrafine-grained microstructure; mechanical strength; fatigue properties

---

1. Introduction

The enhancement in the level of physical and mechanical properties in Al alloys as a result of ultrafine-grained (UFG) structure formation induced by severe plastic deformation (SPD) is currently of great interest [1–5]. As a rule, SPD processing leads to the formation of the microstructure with a grain/subgrain size in the range of 50 nm–1 μm in bulk Al billets [1–5]. Earlier, it was demonstrated that UFG Al alloys have a mechanical strength 1.2–2.0-times as much as their coarse-grained (CG) counterparts processed by a standard aging treatment [1–4,6]. Numerous studies have also shown that other properties, including ductility [1,7,8], fracture resistance [9] and electrical conductivity [1,10,11], can be enhanced. The fatigue behavior of the UFG Al alloys has been widely studied over the last decade, and a detailed overview of those studies can be found in [12–17]. The outcomes of these studies can be roughly generalized as follows. The high-cycle fatigue (HCF) properties are usually not enhanced as significantly as mechanical strength. An insignificant improvement in fatigue life observed in the UFG Al alloys in the HCF regime was related to a low increase in resistance against

Metals **2015**, *5*, 578–590

crack nucleation, as the fatigue life in the HCF is determined by the resistance of material to the crack nucleation [13]. The low cycle fatigue (LCF) behavior of the UFG Al alloys is very complex, since ultrafine grains have a low ability to sustain cyclic loads in the LCF regime. This was related to the limited ductility on monotonic and cycling deformation promoting early crack initiation, as well as a higher fraction of grain boundaries favorable for crack propagation [13]. Unlike CG Al alloys, their UFG counterparts often show an insignificant increase of the stress amplitude (low cyclic hardening) or even its decrease (cyclic softening) when they are cyclically deformed under a constant strain amplitude [18–20]. Nevertheless, intelligent microstructural design in the UFG Al alloys can significantly improve their HCF and LCF behavior [21,22].

It should be noted that the fatigue behavior of the UFG Al alloys has been studied mainly on specimens processed by equal-channel angular pressing (ECAP) [12–14] or by cryorolling [18,23]. It is well known that the ECAP technique or other ECAP-based methods result in the formation of a UFG microstructure with a grain/subgrain size of 300–500 nm [12–14], whereas a microstructure consisting mainly of cells/subgrains having a similar size is formed during cryorolling [18,23]. However, in the literature, there are no reports on the fatigue properties of the UFG Al alloys with a grain size of <200 nm, which can demonstrate record mechanical strength up to 1,000 MPa and are usually processed by HPT [24,25]. The main limitation for fatigue studies on the HPT-processed metallic materials was the small size of the processed disks. Recent progress in up-scaling of the HPT technique along with miniaturization of the specimens for fatigue testing allows one to investigate the fatigue properties of the HPT processed disks [26]. Therefore, the main objective of the present work is to study the fatigue properties and fatigue behavior of an HPT-processed Al-Mg-Si alloy with a grain size of 170 nm. The Al-Mg-Si alloy was selected due to the wide application of the material in construction, auto- and aero-space engineering [21] and as a conductor in electrical engineering [27].

2. Experimental Section

Commercial-grade Al 6061 alloy having a chemical composition (in wt%) 1.00 Mg, 0.61 Si, 0.12 Cu, 0.38 Cr, 0.60 Fe, 0.02 Ti and balance Al was selected for this investigation. Disks having a diameter of 20 mm and a thickness of 1.4 mm were machined from the as-received hot-pressed rod. The disks were solution-treated at 530 °C for 2 h and water quenched. Some disks after water quenching were subjected to the conventional treatment T6, namely artificial ageing at 160 °C for 12 h.

The disks were HPT processed for 10 rotations at RT under a load of 188 tons (e.g., a pressure of 6 GPa). The as-processed disks had a thickness of 0.9 mm. The microstructure and properties of the HPT-processed alloy were studied in the areas located at a distance of ~5 mm (half of radius) from the center of the disks.

Microstructure studies were performed using the transmission electron microscope (TEM, JEOL, Tokyo, Japan) JEM 2100 at an accelerating voltage of 200 kV. Observations were made in both the bright and the dark field imaging modes, and selected area electron diffraction (SAED) patterns were recorded from areas of interest using an aperture of a 1-μm nominal diameter. The lineal intercept method was used to estimate the grain size. At least 200 grains were analyzed to estimate the average grain size.

Specimens for tensile testing and fatigue testing were machined from the solution-treated material and HPT-processed disks. The central axis of the tensile and fatigue specimens was located at a distance of 5.2 mm from the disk center, as shown in Figure 1, where a drawing of the samples is also presented. Two specimens were machined from each disk. All machined tensile and fatigue specimens were mechanically polished to a mirror-like surface using silica colloidal solution at the final stage. Tensile tests were carried out using an Instron 8862 universal testing machine (JEOL, Tokyo, Japan). Tensile specimens were deformed to failure at room temperature with a constant cross-head speed corresponding to an initial strain rate of 10^{-3} s^{-1}. Three tensile specimens were tested for each material condition, and the results thus obtained were found to be reproducible.

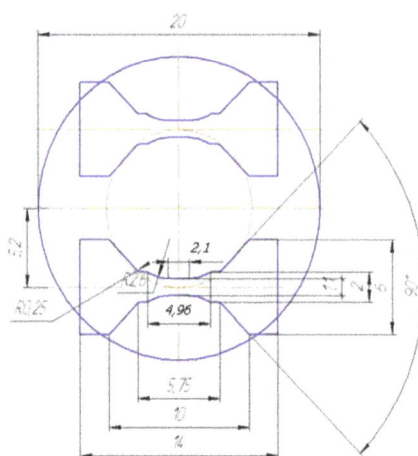

Figure 1. Schematic drawing of specimens machined from disks.

Fatigue testing was carried out using an Instron Electropuls E 3000 testing module with a frequency of 30 Hz under repeated tension conditions with a constant minimum cycle stress of σ_{min} = 100 MPa.

The fatigue fracture surfaces were studied using the scanning electron microscope (SEM, JEOL, Tokyo, Japan) JSM-6490LV operating at an accelerating voltage of 20 kV. The SEM is equipped with the INCA X-sight attachment for energy-dispersive X-ray (EDX) analysis.

3. Results and Discussion

3.1. Effect of HPT Processing on Microstructure of the Al 6061 Alloy

Figure 2 illustrates the microstructure of the Al 6061 alloy before and after HPT processing. Conventional treatment T6 of the as-received alloy resulted in the formation of a CG microstructure with an average grain size of 70 µm (Figure 2a). Particles of primary excess phase forming bands oriented along the pressing direction are observed in the microstructure (Figure 2a). They have a complex chemical composition of $Al_{68}Fe_{21}Si_{6.5}Cr_{2.5}$ (wt%) and are typically formed during the crystallization of slabs. These particles have a spherical shape; their average size is about 1.8 µm, and the volume fraction is about 2.4%. Such a distribution of these particles is usually observed in the hot-pressed Al-Mg-Si alloys. Second phase precipitates having a needle shape are also present in the microstructure (Figure 2b). As is well known, these are the metastable β''-Mg_2Si precipitates, which are typically formed during T6 treatment of the Al-Mg-Si alloys [28,29].

HPT processing of the solution-treated Al 6061 alloy resulted in the formation of a very homogeneous UFG microstructure consisting mainly of equiaxed grains having an average size of 170 nm (Figure 2c,d). This is also confirmed by SAED patterns (Figure 2d). No precipitates, like Guinier–Preston zones and/or second phases, were observed in the processed disks. Such a microstructure is typically formed during HPT processing of the Al-Mg-Si alloys [11,30,31]. It should be also noted that earlier studies using the atom probe tomography (APT) technique revealed segregations of Mg, Cu and Si atoms along grain boundaries in the Al 6061 alloy [30] and in the Al 6060 alloy [32] processed by HPT under similar conditions. HPT processing did not have any effect on the size and volume fraction of the primary excess phases. However, particles changed their orientation and formed a banded structure along the shear plane during HPT processing (Figure 3). It should be noted that in order to avoid the effect of the orientation of the particle bands on the properties of the alloy, specimens for tensile and fatigue testing were cut out, so that particle bands in the gauge sections were oriented along the specimen axis.

3.2. Effect of HPT Processing on Tensile and Fatigue Behavior of the Al 6061 Alloy

The results of mechanical tensile testing of the Al 6061 alloy after conventional T6 treatment and HPT processing are presented in Table 1. It is seen that HPT processing resulted in dramatic improvement of the mechanical strength of the material. The yield strength increased from 276 MPa to 605 MPa, whereas the ultimate tensile strength increased from 365 MPa to 675 MPa. No other SPD processing technique, such as ECAP [33,34], or ECAP combined with cold rolling (CR), or artificial aging (AA) [33], led to such a high mechanical strength in this alloy. Such a significant increase in mechanical strength of the HPT-processed Al 6061 alloy can be related primarily to the formation of a homogeneous UFG structure with a very small grain size, promoting the grain size hardening according to the Hall–Petch relationship [2]. At the same time, segregations of solute atoms in the Al solid solution, which were found earlier in the structure of the HPT-processed Al 6061 alloy in [30], can also provide some additional strengthening.

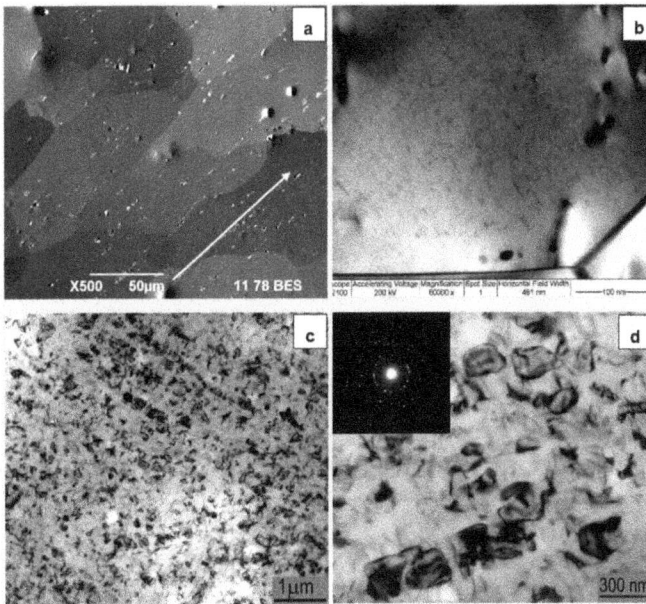

Figure 2. Microstructure of the Al 6061 alloy after conventional T6 treatment (**a**,**b**) and high-pressure torsion (HPT) processing at RT (**c**,**d**). The arrow indicates the pressing direction and the orientation of the bands formed by primary phase particles.

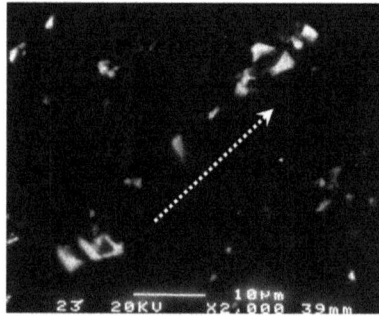

Figure 3. Distribution of the primary excess phases in the Al 6061 alloy processed by HPT at RT (the arrow indicates the shear plane and the orientation of the bands formed by primary phase particles).

Table 1. Mechanical properties of the Al 6061 alloy in the present work in comparison with the mechanical properties reported in earlier studies. ECAP, equal-channel angular pressing; AA, artificial aging; CR, cold rolling; UFG, ultrafine grained; CG, coarse grained.

Processing	State	$\sigma_{0.2}$ (MPa)	σ_{UTS} (MPa)	δ (%)	σ_f (MPa)	σ_f/σ_{UTS}	Reference
T6	CG	276	365	14	100	0.27	present
HPT for 10 turns at RT	UFG	605	675	5.5	200	0.30	work
T6	CG	275	310	12	97	0.31	[34]
ECAP at 125 °C for 1 pass	UFG	310	375	20	80	0.21	[21]
ECAP at 100 °C, 4 passes	UFG	386	434	11	-	-	
ECAP + AA (130 °C for 24 h)	UFG	434	470	10	-	-	[33]
ECAP + CR (15%)	UFG	470	500	8	-	-	

Figure 4 shows the results of fatigue testing of CG T6 treated and HPT-processed material. It is seen that HPT processing has doubled the endurance limit (σ_f) of the material at 10^7 cycles: it increased from 100 MPa in the T6-treated CG alloy to 200 MPa after HPT processing (Figure 4). It should be noted that the σ_f-values obtained after cyclic testing of sub-size specimens with the CG microstructure subjected to the conventional T6 treatment are in very good agreement with the results of the testing of standard reference specimens machined from a similar material [12–14,35]. For example, according to the results of fatigue testing of standard samples, the σ_f-value of the Al 6061 alloy in the T6 state is 97 MPa [13].

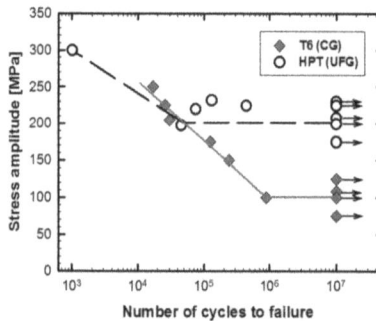

Figure 4. Fatigue curves for the studied Al 6061 alloy.

It should be also noted that the level of σ_f-value achieved in the HPT-processed Al 6061 alloy with an average grain size of 170 nm is also twice the cyclic strength of the UFG material with an average grain size of ~500 nm processed by ECAP in [34].

In Figure 5, the endurance limit is plotted *vs.* the ultimate tensile strength of the CG Al 6061 alloy processed by conventional thermal treatments (O, T4, T6) and by thermo-mechanical treatment (T91) [13,35]. The analysis of the data shows that the ratio of the fatigue limit to the ultimate tensile strength (σ_f/σ_{UTS}) for the given material decreases from ~0.50 to 0.25 with increasing σ_{UTS} (Figure 5). It should be noted that a lower σ_f/σ_{UTS} ratio of 0.20 was reported for the UFG Al 6061 alloy processed by ECAP in [21]. Such low σ_f/σ_{UTS} ratios were found also in other UFG Al-Mg-Si alloys [12–14]. In our work, grain refinement down to 170 nm in the Al 6061 alloy via HPT processing resulted in an increase of the σ_f/σ_{UTS} ratio from 0.27 to 0.30 and in a record value of σ_f = 200 MPa. As is well known, fatigue life in the HCF regime is determined by the resistance of the material to crack initiation [13]. Unlike UFG Al 6061 alloy produced by ECAP in [34], our UFG alloy shows a very homogeneous grain structure and, therefore, should have also a homogeneous resistance to crack nucleation all over the gauge section without any 'weakest link'. Thus, it can be concluded that further grain refinement down to the nanoscale and the generation of a very homogeneous grain structure can be proposed as a recipe for the improvement of HCF properties in the Al alloys.

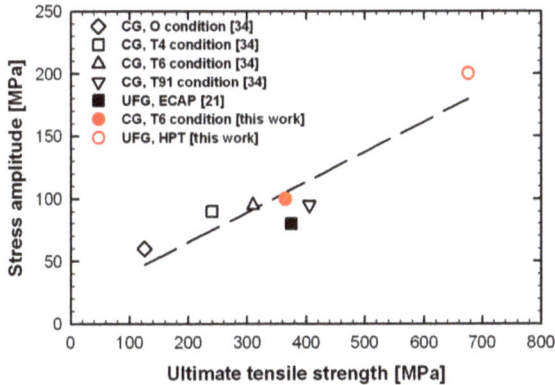

Figure 5. Ultimate tensile strength *vs.* endurance limit for the Al 6061 alloy subjected to various processing treatments.

In the LCF regime, the endurance limit of the HPT-processed Al alloy was somewhat lower than in its CG counterpart (Figure 4). This observation can be rationalized based on the low strain hardening ability of the UFG material, compared to that of the CG T6 one (Table 1), promoting early crack initiation, as well as a much higher volume fraction of grain boundaries favorable for crack propagation [13].

3.3. Analysis of Fatigue Fracture Surfaces

Figures 6 and 7 present SEM images of the fatigue fracture surface of CG T6-treated and HPT-processed Al 6061 alloy, respectively. The area of fatigue crack initiation in the CG T6 material is located on the surface of specimens and is extended into the specimen by ~150 μm (Figure 6a). This is known as Stage I or the short crack growth propagation stage. The increase of the stress intensity factor K as a consequence of crack growth results in the development of slips in different planes close to the crack tip, initiating Stage II, e.g., the stage of stable fatigue crack growth (Figure 6b). An important characteristic of Stage II is the presence of surface ripples, known as "striations". The ductile striations are clearly observed on the fatigue fracture surface (Figure 6c). Their width is in the range of 1–2 µm. The average width of striations is about 3 µm. The most accepted mechanism for the formation of striations is the successive blunting and re-sharpening of the crack tip during cyclic loading [35,36]. Striations are grouped in areas of homogenous propagation separated by tear ridges in which the material fails by shearing, resulting in the apparent continuity of the fatigue advancing front. Finally, Stage III is related to unstable crack growth as K approaches K_{IC} (Figure 6e). At this stage, crack growth is controlled by static modes of failure and is very sensitive to the microstructure, load ratio and stress state. The ductile fracture surface consisting of spherical dimples is observed (Figure 6g). The size of the dimples is in the range of 2–20 µm. Figure 6g shows a fracture pattern within the transition area from a fatigue fracture to the final breakdown.

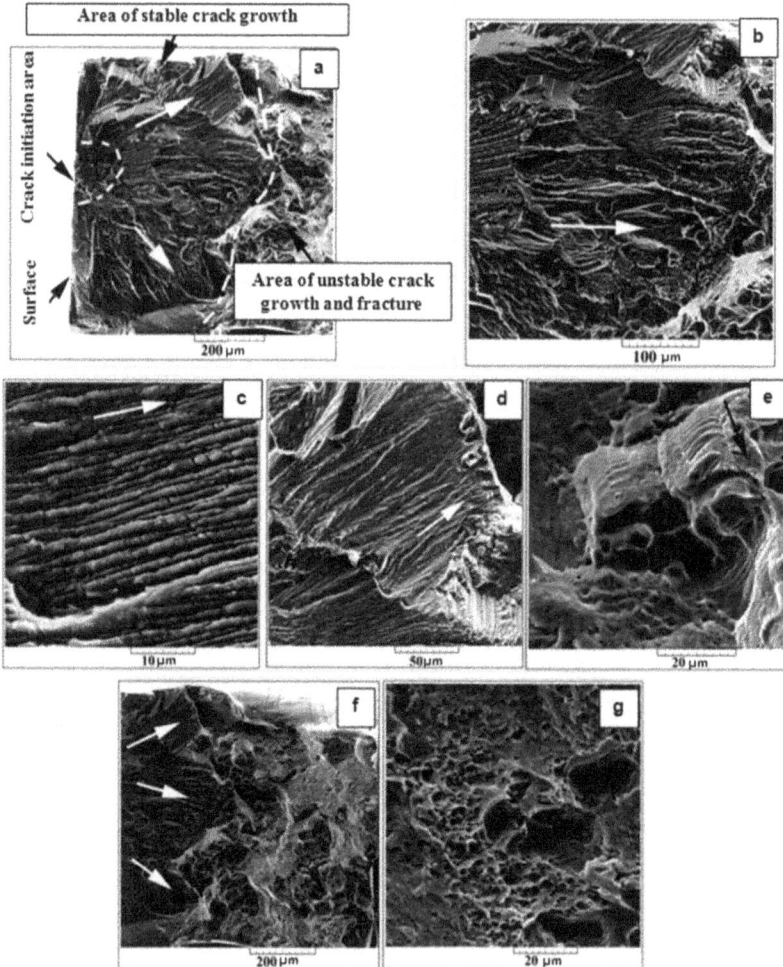

Figure 6. SEM images of the fatigue fracture surface of the CG T6 Al 6061 alloy tested under a stress amplitude of 150 MPa to 2.2 × 10^5 cycles) (the arrows indicate the direction of fatigue crack propagation).

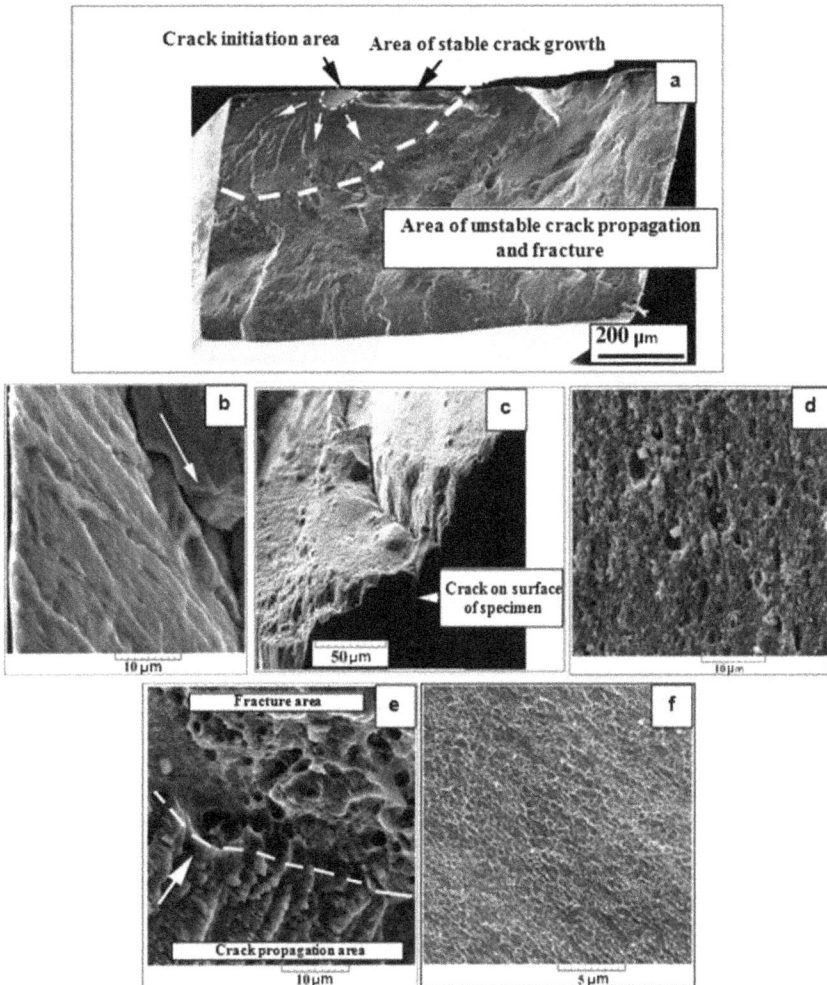

Figure 7. SEM images of the fatigue fracture surface of the UFG Al 6061 alloy: (**a,b,e,f**) tested under a stress amplitude of 225 MPa to 4.4×10^5 cycles; (**c,d**) tested under a stress amplitude of 200 MPa to 4.35×10^4 cycles (the arrows indicate the direction of fatigue crack propagation).

Similar stages of fatigue crack initiation and propagation can be also easily defined on the fatigue fracture surface of the HPT-processed alloy (Figure 7). The area of fatigue crack initiation is also located on the surface of specimens, though its size is smaller: ~50 μm (Figure 7a). The fracture surface relief in the areas of stable and unstable crack growth is less developed as compared to that on in the fatigue-tested CG T6 specimens. Brittle striations are observed in the area of stable crack propagation in the UFG material (Figure 7b) [36]. Unlike in the fatigue-tested CG T6 specimens, stable and unstable crack propagation is accompanied by the formation and growth of micro-cracks in the direction parallel to the applied load (Figure 7c). The length of these micro-cracks is 50–100 μm. These micro-cracks should result in energy dissipation, thus delaying the growth of the main fatigue crack [35]. The dimpled ductile fracture surface is observed in the area of unstable crack growth and the final fracture of the specimen (Figure 7d–g). However, the size of these dimples is in the range of 1–2 μm, which is much lower compared to that in the CG T6 material. A significant difference in the morphology of fatigue fracture surfaces of the CG T6-treated and UFG Al 6061 alloy can be related to the grain structure of these material conditions. As is well known, during the loading phase, the crack is opened by normal stress, which activates plastic slips at the tip. At the crack tip of the CG T6 material with a grain size of 70 μm, dislocations are generated in the grain interior and do not necessarily reach the grain boundaries, when the material is cyclically deformed in Stage II. Conventional transgranular fatigue crack growth followed by specimen failure is observed. In the UFG material with a grain size of 170 nm, dislocation pile-ups are definitely formed at grain boundaries near the crack tip, and intragranular slip coupled with grain boundary sliding can occur [37]. Unaccommodated grain boundary sliding might result in local microcracking or nanovoid formation at grain boundaries. Triple junction nanovoids can also appear due to unaccommodated grain boundary sliding. Both transgranular and intragranular fatigue crack growth can take place in Stage II in this case, though this cannot be clearly resolved in SEM due to the very small grain size. In Stage III, nanovoids can grow and partially relieve the constraints on a grain. Individual single crystal ligaments deform extensively and finally experience chisel-point failure. These grain boundary and triple junction voids also act as sites for nucleation of the dimples in the UFG alloy, which are significantly larger than the individual grains, and the rim of these dimples on the fracture surface (1–2 μm) is typically a magnitude larger than the grain size (170 nm) [37].

4. Conclusions

The following conclusions are outlined.

1. High-pressure torsion (HPT) of the Al 6061 alloy at room temperature leads to the formation of a very homogeneous ultrafine-grained (UFG) microstructure with an average grain size of 170 nm. The yield strength and ultimate tensile strength of the HPT processed alloy are increased as compared to the CG counterpart subjected to the conventional T6 heat treatment from 276 MPa to 605 MPa and from 365 MPa to 675 MPa, correspondingly.

2. HPT processing of the Al 6061 alloy improves its endurance limit by a factor of two (from 100 MPa after T6 treatment to 200 MPa after HPT). This is related to the formation of a very homogeneous UFG microstructure with homogeneous resistance to fatigue crack initiation in the HCF regime. In the LCF regime, the UFG alloy shows somewhat lower fatigue resistance due to its lower strain hardening ability.

3. Classical stages of fatigue crack initiation and propagation are clearly observed on the fatigue fracture surfaces of the CG T6-treated alloy and UFG alloy. Ductile striations are observed in the stage of stable crack propagation in the CG T6-treated alloy, whereas brittle striations seem to dominate on the fatigue fracture surface of the UFG alloy. The dimpled fracture surface is observed at the final stage of crack propagation in both material conditions, with dimples having a smaller size in the UFG alloy.

Acknowledgments: The work has been done under the financial support of the Russian Federal Ministry for Education and Science. R.Z. Valiev and M. Yu. Murashkin gratefully acknowledge the support through Grant No. 14.B25.31.0017, and I. Ovidko acknowledges the support through Programme "5-100-2020". I. Sabirov acknowledges gratefully the Spanish Ministry of Economy and Competitiveness for financial support through the Ramon y Cajal Fellowship.

Author Contributions: Maxim Murashkin formulated the idea of this work, processed the samples, conceived of the workflow and created the initial draft. Ilchat Sabirov carried out mechanical tensile tests and performed microstructural characterization of the processed material. Dmitriy Prosvirnin performed fatigue testing. Vladimir Terentiev performed the SEM studies of the fatigue fracture surfaces. Ilya Ovidko, Ruslan Valiev and Sergey Dobatkin contributed with the overall development of the main concepts presented in this paper. All authors discussed the experimental results, participated in manuscript preparation and approved the final manuscript.

Conflicts of Interest: The authors declare no conflict of interest.

References

1. Valiev, R.Z.; Zhilyaev, A.P.; Langdon, T.G. *Bulk Nanostructured Materials: Fundamentals and Applications*; John Wiley & Sons: Hoboken, NJ, USA, 2014; p. 456.
2. Sabirov, I.; Murashkin, M.Y.; Valiev, R.Z. Nanostructured aluminium alloys produced by severe plastic deformation: New horizons in development. *Mater. Sci. Eng. A* **2013**, *560*, 1–24. [CrossRef]
3. Markushev, M.V.; Murashkin, M.Y. Mechanical properties of submicrocrystalline Al alloys processed by equal-channel angular pressing. *Phys. Met. Metall.* **2000**, *90*, 506–515.
4. Roven, H.J.; Nesboe, H.; Werenskiold, J.C.; Seibert, T. Mechanical properties of aluminum alloys processed by SPD: Comparison of different alloy systems and possible product areas. *Mater. Sci. Eng. A* **2005**, *410–411*, 426–429.
5. Lyakishev, N.P.; Alymov, M.I.; Dobatkin, S.V. Structural bulk nanomaterials. *Russ. Metall.* **2003**, *3*, 191–202.
6. Dobatkin, S.V.; Zakharov, V.V.; Vinogradov, A.Y.; Kitagava, N.; Krasilnikov, N.A.; Rostova, T.D.; Bastrash, E.N. Nanocrystalline structure in Al-Mg-Sc alloys by severe plastic deformation. *Russ. Metall.* **2006**, *6*, 533–540. [CrossRef]
7. Zhao, Y.; Liao, X.; Cheng, S.; Ma, E.; Zhu, Y. Simultaneously increasing the ductility and strength of nanostructured alloys. *Adv. Mater.* **2006**, *18*, 2280–2283. [CrossRef]
8. Horita, Z.; Ohashi, K.; Fujita, T.; Kaneko, K.; Langdon, T.G. Achieving high strength and high ductility in precipitation-hardened alloys. *Adv. Mater.* **2005**, *17*, 1599–1603. [CrossRef]
9. Markushev, M.V.; Bampton, C.C.; Murashkin, M.Y.; Hardwick, D.A. Structure and properties of ultra-fine grained Al alloys produced by severe plastic deformation. *Mater. Sci. Eng. A.* **1997**, *234–236*, 927–931.
10. Murashkin, M.; Sabirov, I.; Kazykhanov, V.; Bobruk, E.; Dubravina, A.; Valiev, R.Z. Enhanced mechanical properties and electrical conductivity in ultra-fine grained Al alloy processed via ECAP-PC. *J. Mater. Sci.* **2013**, *48*, 4501–4509. [CrossRef]
11. Valiev, R.Z.; Murashkin, M.Y.; Sabirov, I. A nanostructural design to produce high-strength Al alloys with enhanced electrical conductivity. *Scr. Mater.* **2014**, *76*, 13–16. [CrossRef]
12. Vinogradov, A.Y.; Khasimoto, S. Fatigue in ultrafine-grained materials processed by equal-channel angular pressing. *Russ. Metall.* **2004**, *1*, 42–51.
13. Estrin, Y.; Vinogradov, A. Fatigue behaviour of light alloys with ultrafine grain structure produced by severe plastic deformation: An overview. *Int. J. Fatigue* **2010**, *32*, 898–907. [CrossRef]
14. Estrin, Y.; Vinogradov, A. Extreme grain refinement by severe plastic deformation: A wealth of challenging science. *Acta Mater.* **2013**, *61*, 782–817. [CrossRef]
15. Höppel, W.; Kautz, M.; Murashkin, M.Y.; Xu, C.; Langdon, T.G.; Valiev, R.Z.; Mughrabi, H. An overview: Fatigue Behavior of Ultrafine-Grained Metals and Alloys. *Int. J. Fatigue* **2006**, *28*, 1001–1010. [CrossRef]
16. Höpel, H.W.; Göken, M. Fatigue Behavior in Nanostructured Metals. In *Nanostructured Metals and Alloys: Processing, Microstructure, Mechanical Properties and Applications*; Whang, S.H., Ed.; Woodhead Publishing Limited: Suite, PA, USA, 2011; pp. 507–541.
17. Mughrabi, H.; Höppel, H.W.; Kautz, M. Fatigue and microstructure of ultra-fine grained metals produced by severe plastic deformation. *Scr. Mater.* **2004**, *51*, 807–812. [CrossRef]
18. Malekjani, S.; Hodgson, P.D.; Cizek, P.; Sabirov, I.; Hilditch, T.B. Cyclic deformation response of UFG 2024 Al alloy. *Int. J. Fatigue* **2011**, *33*, 700–709. [CrossRef]
19. Canadinc, D.; Maier, H.J.; Gabor, P.; May, J. On the cyclic deformation response of ultrafine-grained Al–Mg alloys at elevated temperatures. *Mater. Sci. Eng. A* **2008**, *496*, 114–120. [CrossRef]
20. Vinogradov, A.; Washikita, A.; Kitagawa, K.; Kopylov, V.I. Fatigue life of fine-grain Al-Mg-Sc alloys produced by equal-channel angular pressing. *Mater. Sci. Eng. A* **2003**, *349*, 318–326. [CrossRef]
21. Chung, C.S.; Kim, J.K.; Kim, H.K.; Kim, W.J. Improvement of high-cycle fatigue life in a 6061 Al alloy produced by equal channel angular pressing. *Mater. Sci. Eng. A* **2002**, *337*, 39–44. [CrossRef]
22. Lapovok, R.; Loader, C.; Dalla Torre, F.H.; Semiatin, S.L. Microstructure evolution and fatigue behavior of 2124 aluminum processed by ECAE with back pressure. *Mater. Sci. Eng. A* **2006**, *425*, 36–46. [CrossRef]
23. Malekjani, S.; Hodgson, P.D.; Cizek, P.; Hilditch, T.B. Cyclic deformation response of ultra-fine pure Al. *Acta Mater.* **2011**, *59*, 5358–5367. [CrossRef]
24. Valiev, R.Z.; Enikeev, N.A.; Murashkin, M.Y.; Kazykhanov, V.U.; Sauvage, X. On the origin of the extremely high strength of ultra-fine grained Al alloys. *Scr. Mater.* **2010**, *63*, 949–952. [CrossRef]

25. Liddicoat, P.V.; Liao, X.Z.; Zhao, Y.; Zhu, Y.; Murashkin, M.Y.; Lavernia, E.J.; Valiev, R.Z.; Ringer, S.P. Nanostructural hierarchy increases the strength of aluminium alloys. *Nat. Comm.* **2010**, *1*, 63–70. [CrossRef]

26. Ruffing, C.; Ivanisenko, Y.; Kerscher, E. Fatigue behavior of ultrafine grained medium Carbon steel processed by severe plastic deformation. *IOP Conf. Series: Mater. Sci. Eng.* **2014**, *63*, 012163. [CrossRef]

27. Polmear, I.J. Light Alloys. In *From Traditional Alloys to Nanocrystals*; Butterworth-Heinemann: Oxford, UK, 2006; p. 417.

28. Mondolfo, L.F. *Aluminum Alloys: Structure and Properties*; Butterworth: Oxford, UK, 1976; p. 971.

29. Murayama, M.; Hono, K. Pre-precipitate clusters and precipitation processes in Al-Mg-Si alloys. *Acta Mater.* **1999**, *47*, 1537–1548. [CrossRef]

30. Nurislamova, G.; Sauvage, X.; Murashkin, M.; Islamgaliev, R.; Valiev, R. Nanostructure and related mechanical properties of an Al-Mg-Si alloy processed by severe plastic deformation. *Phil. Mag. Lett.* **2008**, *88*, 459–466. [CrossRef]

31. Moreno-Valle, E.C.; Sabirov, I.; Perez-Prado, M.T.; Murashkin, M.Y.; Bobruk, E.V.; Valiev, R.Z. Effect of the grain refinement via severe plastic deformation on strength properties and deformation behavior of an Al 6061 alloy at room and cryogenic temperatures. *Mater. Lett.* **2011**, *65*, 2917–2919. [CrossRef]

32. Sha, G.; Tugcu, K.; Liao, X.Z.; Trimby, P.W.; Murashkin, M.Y.; Valiev, R.Z.; Ringer, S.P. Strength, grain refinement and solute nanostructures of an Al-Mg-Si alloy (AA6060) processed by high-pressure torsion. *Acta Mater.* **2014**, *63*, 169–179. [CrossRef]

33. Murashkin, M.Y.; Markushev, M.V.; Ivanisenko, Y.V.; Valiev, R.Z. Strength of commercial aluminum alloys after equal channel angular pressing and post-ECAP processing. *Solid State Phen.* **2006**, *114*, 91–96. [CrossRef]

34. Hatch, J.E. *Aluminium: Properties and Physical Metallurgy*; ASM International: Materials Park, OH, USA, 1984; p. 636.

35. Suresh, S. *Fatigue of Materials*; Cambridge University Press: Cambridge, UK, 1998; p. 679.

36. Milella, P.P. *Fatigue and Corrosion in Metals*; Springer: Berlin, Germany, 2013; p. 763.

37. Dalla Torre, F.; van Swygenhoven, H.; Victoria, M. Nanocrystalline electrodeposited Ni: Microstructure and tensile properties. *Acta Mater.* **2002**, *50*, 3957–3970. [CrossRef]

metals

MDPI

Article

Influence of Particulate Reinforcement and Equal-Channel Angular Pressing on Fatigue Crack Growth of an Aluminum Alloy

Lisa Köhler *, Kristin Hockauf and Thomas Lampke

Institute of Materials Science and Engineering, Technische Universität Chemnitz, Erfenschlager Str. 73, 09125 Chemnitz, Germany; kristin.hockauf@mb.tu-chemnitz.de (K.H.); thomas.lampke@mb.tu-chemnitz.de (T.L.)

* Author to whom correspondence should be addressed; lisa.koehler@mb.tu-chemnitz.de; Tel.: +49-371-531-32632; Fax: +49-371-531-23819.

Academic Editor: Heinz Werner Höppel

Received: 26 February 2015; Accepted: 12 May 2015; Published: 18 May 2015

Abstract: The fatigue crack growth behavior of unreinforced and particulate reinforced Al 2017 alloy, manufactured by powder metallurgy and additional equal-channel angular pressing (ECAP), is investigated. The reinforcement was done with 5 vol % Al_2O_3 particles with a size fraction of 0.2–2 µm. Our study presents the characterization of these materials by electron microscopy, tensile testing, and fatigue crack growth measurements. Whereas particulate reinforcement leads to a drastic decrease of the grain size, the influence of ECAP processing on the grain size is minor. Both reinforced conditions, with and without additional ECAP processing, exhibit reduced fatigue crack growth thresholds as compared to the matrix material. These results can be ascribed to the well-known effect of the grain size on the crack growth, since crack deflection and closure are directly affected. Despite their small grain size, the thresholds of both reinforced conditions depend strongly on the load ratio: tests at high load ratios reduce the fatigue threshold significantly. It is suggested that the strength of the particle-matrix-interface becomes the critical factor here and that the particle fracture at the interfaces dominates the failure behavior.

Keywords: Al 2017 alloy; Al_2O_3 particulate reinforcement; fatigue crack growth; equal-channel angular pressing (ECAP)

1. Introduction

Aluminum matrix composites (AMCs) are designed to meet the requirements of advanced engineering applications. Reinforcement with ceramic particles provides high strength and stiffness, and enhances wear resistance, thermal stability, and creep resistance [1,2]. It has been shown that methods of severe plastic deformation (SPD) can help to homogenize the particle distribution [3], which improves the mechanical properties. Equal-channel angular pressing (ECAP) has emerged as one of the most frequently used SPD methods for processing particulate reinforced AMCs [4–7]. For unreinforced materials, ECAP processing leads to strain-hardening of the material and decreases the grain size by the rearrangement of dislocations to subgrain boundaries and cell walls [8–10].

Focusing on potential future applications for AMCs produced by ECAP, fatigue crack growth is an important field of interest. To the best of our knowledge, previous studies only focused on crack propagation either in ultrafine-grained materials produced by ECAP or in AMCs. These investigations show the well-known effect of the grain or particle size on the fatigue crack growth. The grain refinement through ECAP leads to a minimized amount of crack deflection and roughness-induced crack closure, which results in higher crack propagation rates and, therefore, in minor thresholds [11–18]. In reinforced materials, particle sizes larger than 2–5 µm have a contrary effect on crack propagation: by increasing crack deflection and roughness-induced crack closure, crack propagation rates are lowered [19–24].

The purpose of the present study is to investigate the influence of particulate reinforcement with a particle size less than 2 µm and additional ECAP on the near-threshold fatigue crack growth of an aluminum alloy. This small particle size was chosen in order to minimize local stress concentrations in the matrix material [25]. It is well-known that small particles cause significantly higher strain hardening and that they are less susceptible to cracking as compared to coarser particles [26,27]. As matrix material, the Al 2017 alloy has been chosen here and in preliminary studies [28], because it combines attractive properties, such as good fracture toughness in the underaged condition, high resistance against crack propagation, and excellent high-temperature strength, which enables future applications for lightweight structures, e.g., for the aerospace and aeronautical industries.

2. Experimental Section

2.1. Material

A commercially available, gas-atomized, spherical Al 2017 powder alloy was used as matrix material. The particle size fraction of the powder, with a chemical composition such as that presented in Table 1, was below 100 µm. Al_2O_3 powder with a particle size fraction of 0.2–2 µm was used as reinforcing component. The composite powder, consisting of 95 vol % matrix material and 5 vol % Al_2O_3 particles, was processed in a high-energy ball mill and hot-degassed at 450 °C and 0.06 bar for 4 h. Afterwards, compaction was performed by hot isostatic pressing at 450 °C and 1100 bar for 3 h. The manufacturing process is described in more detail in [28].

Table 1. Chemical composition of the Al 2017 powder alloy.

Element	Al	Cu	Mn	Mg	Fe	Si	Zn
wt %	>94.1	4.1	0.8	0.6	0.2	0.1	traces

The non-ECAP-processed conditions, unreinforced, and reinforced, were solid-solution treated at 505 °C for 60 min, water-quenched, and subsequently naturally aged at room temperature for one month in order to adjust an underaged condition. For the ECAP-processed conditions, as described in [28], a pre-ECAP aging at 140 °C was carried out for 10 min to increase the workability. Immediately after the pre-ECAP aging, the material was ECAP-processed at 140 °C for one pass. The ECAP processing was performed in a device with an internal angle of 120 and a cross-section of 15 × 15 mm², equipped with movable walls and a bottom slider. This processing resulted in an equivalent strain of 0.67 in each passage [29]. After ECAP, the material was peak-aged at 140 °C for 360 min to optimize the ductility. With this combined treatment of ECAP and aging, both elevated strength and moderate ductility can be achieved [30,31].

2.2. Methods of Mechanical Testing and Electron Microscopy

Quasi-static tensile tests were performed in a Zwick-Roell servohydraulic testing machine (Zwick, Ulm, Germany) at a strain rate of $10^{-3} \cdot s^{-1}$ at room temperature. Cylindrical specimens were used with a cross section of 3.5 mm and a gauge length of 10.5 mm.

Fatigue crack propagation measurements were performed ΔK-controlled in a Rumul testronic resonant testing machine (Russenberger Prüfmaschinen AG, Neuhausen am Rheinfall, Switzerland) at room temperature and at three constant load ratios: $R = 0.1$, 0.4 and 0.7. Single-edge-notched bend (SEB) specimens with a geometry according to ASTM E399-12e3 [32] and a thickness of 5 mm were used. Figure 1 shows the position in the ECAP billet out of which the SEB specimens were extracted. In the choice of the extraction direction, it was considered that the crack is most likely sidetracked in the direction of the last ECAP shear plane [12]. The crack length was measured continuously according to the principle of the indirect electric potential method with crack measurement foils.

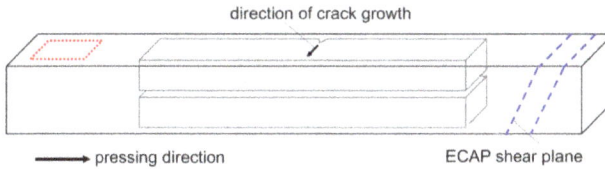

Figure 1. Position of two single-edge-notched bend specimens within the ECAP billet. The ideal direction of crack growth is parallel to the ECAP shear plane. The plane from which samples for microstructural analyses were taken is marked in red.

After precracking at $\Delta K = 2–7$ MPam$^{1/2}$ (depending on material condition and load ratio), ΔK was lowered in small steps until the fatigue threshold ΔK_{th} was reached. Then, 0.5 mm crack growth were realized at this cyclic stress intensity to avoid any load history effect from the threshold region. In the following, ΔK was raised again in small steps up to a cyclic stress intensity of about 11 MPam$^{1/2}$ and a crack length of 5 mm.

Samples for microstructural analyses were extracted parallel to the extrusion direction, as shown in Figure 1, and electro-polished followed by a final ion milling polishing step. These samples were analyzed by scanning transmission electron microscopy (STEM) and quadrant back scatter diffraction (QBSD) at an accelerating voltage of 30 kV, as well as by electron back scatter diffraction (EBSD, EDAX TSL OIM 5.2, AMETEK GmbH, Wiesbaden, Germany) at 15 kV using a Zeiss Neon 40 field emission microscope (Carl Zeiss MicroImaging GmbH, Jena, Germany).

3. Results and Discussion

3.1. Tensile Tests

The tensile properties of the investigated material conditions are shown in Table 2. The unreinforced aluminum matrix (in an underaged heat treatment condition) exhibits the lowest yield strength with 252 MPa just as the highest ductility, which shows in a uniform elongation of 20%. ECAP processing (with subsequent aging as described in the experimental section) leads to a significant increase in yield strength (487 MPa, which is an increase of 93%) in combination with a moderate ductility (11% uniform elongation). The strategy of combining ECAP and aging, which is used here and which is referred to as "optimizing heat treatment" in [33], enables this favorable combination of strength and ductility: ECAP processing results in a severely strain hardened material with a high dislocation density. In the following process of thermal aging, numerous finely-dispersed precipitates form and strengthen the matrix. Simultaneously, thermal recovery takes place and increases the material's ductility.

Table 2. Tensile properties of the unreinforced and reinforced conditions.

Material	Yield Strength in MPa	Ultimate Tensile Strength in MPa	Uniform Elongation in %	Elongation to Fracture in %
Al 2017 unreinforced	252	435	20	29
Al 2017 unreinforced after 1 ECAP pass	487	566	11	17
Al 2017 with 5 vol % Al$_2$O$_3$	328	511	11	13
Al 2017 with 5 vol % Al$_2$O$_3$ after 1 ECAP pass	509	580	5	8

The dispersion strengthening, which is caused by the reinforcement with 5 vol % particles, leads to an improvement in strength, which is less compared to the strengthening, which is achieved by ECAP of the unreinforced material. ECAP processing of the particulate reinforced material increases the yield strength from 328 MPa to 509 MPa (which is an increase of 55%). This shows, that the effect of strain hardening during ECAP is less pronounced than in the unreinforced material. Both reinforced conditions exhibit minor ductility, which is ascribed to the particles which act as interior notches [28].

3.2. Microstructure

One ECAP pass of the unreinforced matrix material leads to a bimodal grain structure with inhomogeneously distributed refined grains. The initial grain size is significantly reduced in the shear bands, which is shown in Figure 2. In the outside parts, almost no grain refinement has taken place.

The reinforced material exhibits elongated areas with a width of 100–200 μm and a height of about 7 μm without reinforcement components. These areas are preferentially orientated parallel to the direction of extrusion. In Figure 3, micrographs of the reinforced areas are shown. The Al_2O_3 particles are finely dispersed and exhibit an intact interface to the aluminum matrix. In these micrographs, the presence of the coarse intermetallic phase Al_2Cu can also be seen. Additional ECAP does not lead to significant grain refinement (see Figure 4).

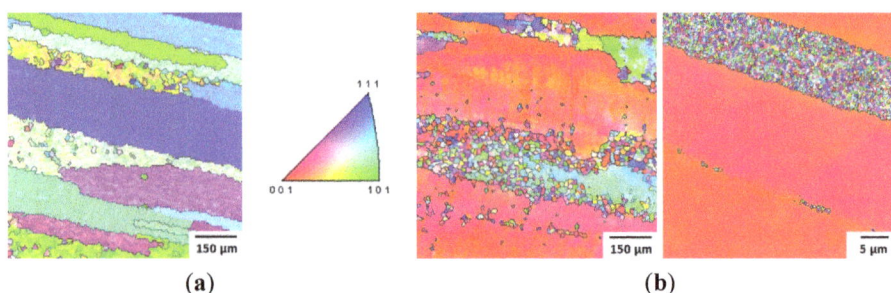

(a) (b)

Figure 2. Electron back scatter diffraction (EBSD) color-coded orientation maps of the unreinforced matrix material conditions: (**a**) non-ECAP-processed and (**b**) after one ECAP pass. In the shear bands, the grain size is considerably reduced after one ECAP pass. By contrast, the initial grain size outside the shear bands is almost maintained.

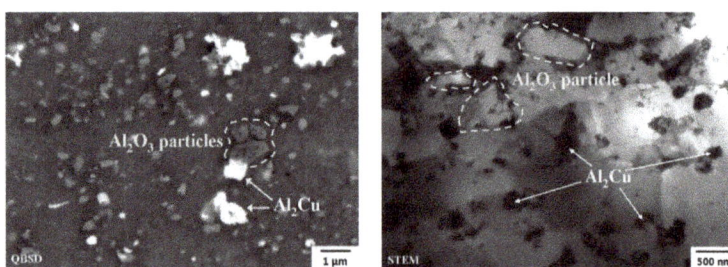

Figure 3. Quadrant backscatter diffraction (QBSD, **left**) and scanning transmission electron microscopy (STEM, **right**) micrographs of Al 2017 with 5 vol % Al_2O_3 particles with a size of 0.2–2 μm. Al_2O_3 particles are marked with broken lines and the Al_2Cu precipitates with arrows. The particles are finely dispersed and matrix and particles exhibit an intact interface.

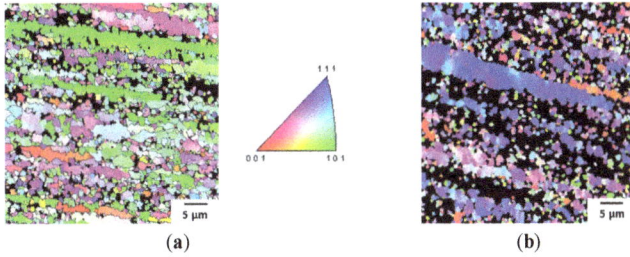

Figure 4. EBSD color-coded orientation maps of the reinforced conditions: (**a**) non-ECAP-processed and (**b**) after additional ECAP. For the reinforced material, ECAP does not lead to a significant grain refinement.

3.3. Fatigue Crack Surface Analysis

For the unreinforced conditions the fatigue crack surfaces are shown in Figure 5. The non-ECAP-processed condition exhibits a strongly faceted surface with pronounced band-typed steps. This fissured crack surface can be explained by the large grain size of this condition and thus by a high amount of crack deflection. For the material after one ECAP pass, plateaus are present on the crack surface. Additionally, striations are clearly visible. Such signs of "micro-damage" have also been observed for other ECAP-processed aluminum alloys [16]. They were interpreted as the severely strain-hardened material is not being able to compensate the stresses at the crack tip by deforming plastically. Compared to the unprocessed material, the crack path after ECAP is less tortuous and more straight-forward. Presumably, the crack propagates preferably along or within the shear bands.

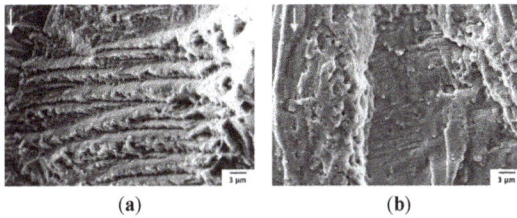

Figure 5. Fatigue crack surfaces of the unreinforced conditions: (**a**) non-ECAP-processed and (**b**) after one ECAP pass near the fatigue threshold. The crack propagation direction is marked with an arrow. The surface of the non-ECAP-processed condition exhibits a band-type faceted appearance, whereas ECAP leads to plateaus and clearly visible striations.

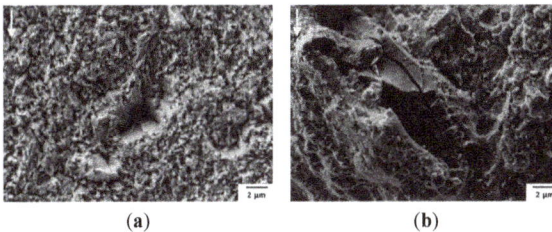

Figure 6. Fatigue crack surfaces of the reinforced conditions: (**a**) non-ECAP-processed and (**b**) after one ECAP pass near the fatigue threshold. The crack propagation direction is marked with an arrow. The surface is less rough and barely fissured as compared to the unreinforced conditions. Through additional ECAP, the particle failure is more likely to fracture than to break out.

In Figure 6, the fatigue crack surfaces of the reinforced conditions are shown. In contrast to the unreinforced conditions, the crack surfaces are characterized by a minor roughness. Hardly any faceting is noticeable. Due to the small grain and particle sizes of these conditions, crack deflection is reduced to a minimum. This outcome is in good agreement with the findings in [19,34]. It is noticeable that, in the ECAP-processed condition, particle fracture is more pronounced, whereas the initial reinforced condition exhibits a higher rate of particle break outs. It cannot be concluded with certainty if the particle fracture proceeded due to fatigue loading or whether it was caused by ECAP.

3.4. Fatigue Crack Propagation

Fatigue crack growth curves near the threshold for the tested load ratios are shown in Figure 7. The matrix material exhibits the highest thresholds and the lowest fatigue crack growth rates for all tested conditions. Due to the major grain size of the matrix material, the amount of crack deflection and roughness-induced crack closure and, therefore, the strongly branched crack path lead to decreased fatigue crack growth rates, which results in major thresholds. One ECAP pass results in higher crack growth rates and lower thresholds. Presumably, due to the preferred crack propagation along the shear bands and the finer grains in this area, less crack deflection and closure occurs. Supposedly, also the high dislocation density induced by ECAP inhibits stress relief and, therefore, early damaging at the crack tip takes place [13].

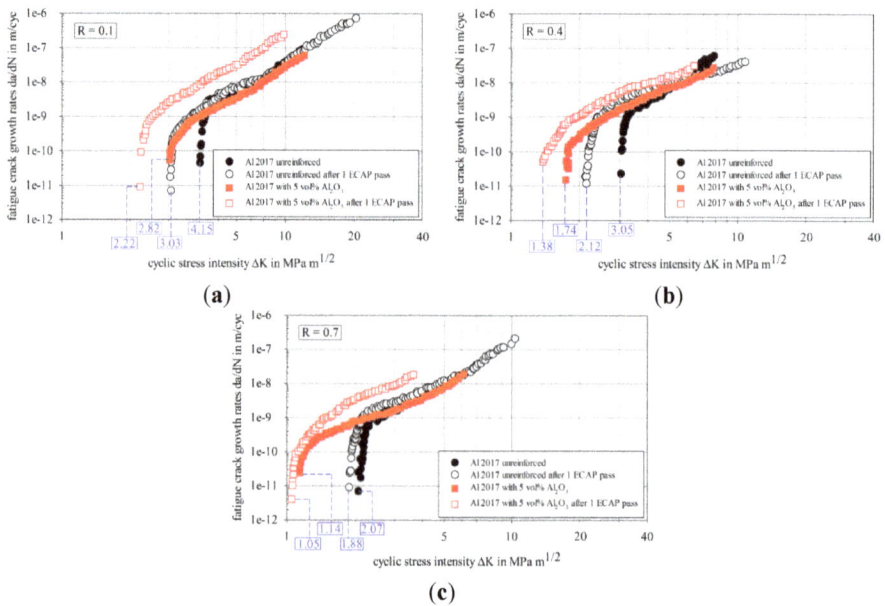

Figure 7. Fatigue crack growth curves of Al 2017 in unreinforced and reinforced conditions with additional ECAP at load ratios of: (**a**) *R* = 0.1, (**b**) *R* = 0.4, and (**c**) *R* = 0.7. The thresholds for the reinforced conditions are lower than for the matrix material conditions. With the increase of the load ratio, this effect becomes more pronounced. The crack growth rates for the reinforced ECAP-processed conditions are generally the highest.

For the reinforced conditions, at low cyclic stress intensities, the crack propagation rates are higher as compared to the matrix material conditions. As the near threshold regime is highly sensitive to the microstructure, the main factor for this behavior is presumably the significantly smaller grain size and the resulting minimized effects of crack deflection and roughness-induced crack closure. For the

ECAP-processed conditions, fatigue crack growth thresholds are further reduced as compared to their corresponding not-processed conditions. This is probably due to the high dislocation density in these materials, which leads to a diminished deformation ability at the crack tip.

Figure 8 presents the fatigue threshold depending on the load ratio for all tested conditions. The matrix material and both reinforced conditions exhibit a strong dependency on the load ratio. Although the curve progression is nearly parallel for these three conditions, the underlying mechanisms are different. The matrix material shows the highest thresholds and also a strong load ratio dependency. This can be explained by the well-known effect of the grain size on the crack growth at different load ratios [11–13,15,35]. The amount of crack deflection is constant for all load ratios, whereas the effect of roughness induced crack closure is minimized by increasing the load ratio. The fatigue thresholds of the ECAP-processed matrix material show a minor dependency on the load ratio. Supposedly, the small grain size in the shear bands and therefore the generally low amount of crack deflection and closure lead here to minimized load ratio effects.

However, despite their small grain size, both reinforced conditions exhibit a strong dependency on the load ratio. In this case, assumedly, the thresholds depend critically on the interaction between the softer aluminum matrix where crack propagation takes place and the interface where decohesion or brittle particle facture can occur [36]. At higher load ratios, the susceptibility to cracking of the particle-matrix-interface is strongly increased. On that account, fatigue thresholds further decrease for higher load ratios. This, in turn, results in a higher load-ratio-dependency. For the reinforced condition, additional ECAP does not influence the load ratio dependency.

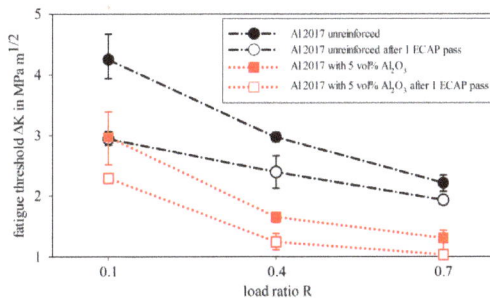

Figure 8. Fatigue thresholds of Al 2017 in unreinforced and reinforced conditions with additional ECAP at different load ratios. The ECAP-processed matrix material exhibits the least dependency on the load ratio. In contrast, the matrix material and the reinforced conditions depend strongly on the load ratio and display almost the same amount for the three conditions.

4. Conclusions

In this paper, the effect of particulate reinforcement and equal-channel angular pressing on the fatigue crack growth in an aluminum alloy produced by powder metallurgy is investigated. For the Al 2017 powder alloy, reinforced with 5 vol % Al_2O_3 with a size fraction of 0.2–2 µm, the fatigue properties are compared to the unreinforced material. Two different conditions for both materials are chosen: non-processed and after one pass of ECAP.

(1) One ECAP pass of the unreinforced material results in an inhomogeneous microstructure with shear bands. These shear bands are characterized by a considerably reduced grain size. In the surrounding areas, the grains are strongly elongated, but not refined. The grain size of the reinforced material is significantly lower, as compared to the unreinforced. Additional ECAP does not lead to a significant grain refinement, here.

(2) The yield strength of both materials, unreinforced and reinforced, is increased by one pass of ECAP. However, the strain hardening effect of ECAP is more pronounced in the unreinforced material.

(3) Due to the major grain size of the unreinforced matrix material, the high amount of crack deflection and roughness-induced crack closure lead to a tortuous crack path and therefore to minor crack propagation rates. The grain refinement through one ECAP pass reduces these effects. Supposedly because crack propagation occurred in the finer-grained shear bands, the fatigue thresholds and their dependency on the load ratio are diminished for the unreinforced material after ECAP.

(4) For the reinforced material, the fatigue thresholds are drastically reduced due to the significantly smaller grain size of these conditions. Despite their small grain size, the reinforced materials exhibit a strong dependency on the load ratio. Here, the interaction between the aluminum matrix and the interface is perceived as crucial: the susceptibility to cracking of the particle-matrix-interface is strongly increased at higher load ratios. This leads to a "particle-failure-induced" load ratio dependency.

Acknowledgments: The authors gratefully acknowledge funding of the Collaborative Research Center SFB 692 received from the German Research Foundation (Deutsche Forschungsgemeinschaft DFG).

Author Contributions: L.K. performed and analyzed the experiments and is the primary author of the paper. K.H. conceived and designed the experiments and discussed the results and analysis with the author. T.L. supervised the work.

Conflicts of Interest: The authors declare no conflict of interest.

References

1. Chawla, N.; Shen, Y.L. Mechanical behavior of particle reinforced metal matrix composites. *Adv. Eng. Mater.* **2001**, *3*, 357–370. [CrossRef]
2. Mani, B.; Paydar, M.H. Application of forward extrusion-equal channel angular pressing (FE-ECAP) in fabrication of aluminum metal matrix composites. *J. Alloys Compd.* **2010**, *492*, 116–121. [CrossRef]
3. Tan, M.J.; Zhang, X. Powder metal matrix composites: Selection and Processing. *Mater. Sci. Eng. A* **1998**, *244*, 80–85. [CrossRef]
4. Ramu, G.; Bauri, R. Effect of equal channel angular pressing (ECAP) on microstructure and properties of Al-SiC$_p$ composites. *Mater. Des.* **2009**, *30*, 3554–3559. [CrossRef]
5. Sabirov, I.; Kolednik, O.; Valiev, R.Z.; Pippan, R. Equal channel angular pressing of metal matrix composites: Effect on particle distribution and fracture toughness. *Acta Mater.* **2005**, *53*, 4919–4930. [CrossRef]
6. Valiev, R.Z.; Islamgaliev, R.K.; Kuzmina, N.F.; Li, Y.; Langdon, T.G. Strengthening and grain refinement in an Al-6061 metal matrix composite through intense plastic straining. *Scr. Mater.* **1999**, *40*, 117–122. [CrossRef]
7. Muñoz-Morris, M.A.; Calderón, N.; Guitierrez-Urrutia, I.; Morris, D.G. Matrix grain refinement in Al-TiAl composites by severe plastic deformation: Influence of particle size and processing route. *Mater. Sci. Eng. A* **2006**, *425*, 131–137. [CrossRef]
8. Segal, V.M.; Reznikov, V.I.; Drobyshevskiy, A.E.; Kopylov, V.I. Plastic working of metals by simple shear (Russian translation). *Russ. Metall.* **1981**, *1*, 99–105.
9. Valiev, R.Z.; Langdon, T.G. Principles of equal-channel pressing as a processing tool for grain refinement. *Prog. Mater. Sci.* **2006**, *51*, 881–981. [CrossRef]
10. Furukawa, M.; Horita, Z.; Langdon, T.G. Developing ultrafine grain sizes using severe plastic deformation. *Adv. Eng. Mater.* **2001**, *3*, 121–125. [CrossRef]
11. Pao, P.S.; Jones, H.N.; Cheng, S.F.; Feng, C.R. Fatigue crack propagation in ultrafine grained Al-Mg alloy. *Int. J. Fatigue* **2005**, *27*, 1164–1169. [CrossRef]
12. Hübner, P.; Kiessling, R.; Biermann, H.; Hinkel, T.; Jungnickel, W.; Kawalla, R.; Höppel, H.-W.; May, J. Static and cyclic crack growth behavior of ultrafine-grained Al produced by different severe plastic deformation methods. *Metall. Mater. Trans. A* **2007**, *38*, 1926–1933. [CrossRef]

13. Vinogradov, A. Fatigue limit and crack growth in ultra-fine grain metals produced by severe plastic deformation. *J. Mater. Sci.* **2007**, *42*, 1797–1808. [CrossRef]
14. Cavaliere, P. Fatigue properties and crack behavior of ultra-fine and nanocrystalline pure metals. *Int. J. Fatigue* **2009**, *31*, 1476–1489. [CrossRef]
15. Estrin, Y.; Vinogradov, A. Fatigue behaviour of light alloys with ultrafine grain structure produced by severe plastic deformation: An overview. *Int. J. Fatigue* **2010**, *32*, 898–907. [CrossRef]
16. Meyer, L.W.; Sommer, K.; Halle, T.; Hockauf, M. Crack growth in ultrafine-grained AA6063 produced by equal-channel angular pressing. *J. Mater. Sci.* **2008**, *43*, 7426–7431. [CrossRef]
17. Avtokratova, E.V.; Sitdikov, O.S.; Kaibyshev, R.O.; Watanabe, Y. Behavior of a submicrocrystalline aluminum alloy 1570 under conditions of cyclic loading. *Phys. Met. Metallogr.* **2009**, *107*, 291–297. [CrossRef]
18. Hockauf, K.; Wagner, M.F.-X.; Halle, T.; Niendorf, T.; Hockauf, M.; Lampke, T. Influence of precipitates on low-cycle fatigue and crack growth behavior in an ultrafine-grained aluminum alloy. *Acta Mater.* **2014**, *80*, 250–263. [CrossRef]
19. Chen, Z.Z.; Tokaji, K. Effects of particle size on fatigue crack initiation and small crack growth in SiC particulate-reinforced aluminum alloy composites. *Mater. Lett.* **2004**, *58*, 2314–2321. [CrossRef]
20. Levin, M.; Karlsson, B. Influence of SiC particle distribution and prestraining on fatigue crack growth rates in aluminum AA 6061-SiC composite material. *Mater. Sci. Technol.* **1991**, *7*, 596–607. [CrossRef]
21. Mason, J.J.; Ritchie, R.O. Fatigue crack growth resistance in SiC particulate and whisker reinforced P/M 2124 aluminum matrix composites. *Mater. Sci. Eng. A* **1997**, *231*, 170–182. [CrossRef]
22. Llorca, J. Fatigue of particle- and whisker-reinforced metal-matrix composites. *Prog. Mater. Sci.* **2002**, *47*, 283–353. [CrossRef]
23. Oh, K.H.; Han, K.S. Short-fiber/ particle hybrid reinforcement: Effects on fracture toughness and fatigue crack growth of metal matrix composites. *Comput. Sci. Technol.* **2007**, *67*, 1719–1726. [CrossRef]
24. Chawla, N.; Ganesh, V.V. Fatigue crack growth of SiC particle reinforced metal matrix composites. *Int. J. Fatigue* **2010**, *32*, 856–863. [CrossRef]
25. Huang, M.; Li, Z. Size effects on stress concentration induced by a prolate ellipsoidal particle and void nucleation mechanism. *Int. J. Plast.* **2005**, *21*, 1568–1590. [CrossRef]
26. Xue, Z.; Huang, Y.; Li, M. Particle size effect in metallic materials: A study by the theory of mechanism-based strain gradient plasticity. *Acta Mater.* **2002**, *50*, 149–160. [CrossRef]
27. Slipenyuk, A.; Kuprin, V.; Milman, Y.; Goncharuk, V.; Eckert, J. Properties of P/M processed particle reinforced metal matrix composites specified by the reinforcement concentration and matrix-to-reinforcement particle size ratio. *Acta Mater.* **2006**, *54*, 157–166. [CrossRef]
28. Wagner, S.; Siebeck, S.; Hockauf, M.; Nestler, D.; Podlesak, H.; Wielage, B.; Wagner, M.F.X. Effect of SiC-reinforcement and equal-channel angular pressing on microstructure and mechanical properties of AA2017. *Adv. Eng. Mater.* **2012**, *14*, 388–393. [CrossRef]
29. Iwahashi, Y.; Wang, J.T.; Horita, Z.; Nemoto, M.; Langdon, T.G. Principle of equal-channel angular pressing for the processing of ultra-fine grained materials. *Scr. Mater.* **1996**, *35*, 143–146. [CrossRef]
30. Kim, J.K.; Jeong, H.G.; Hong, S.I.; Kim, Y.S.; Kim, W.J. Effect of aging treatment on heavily deformed microstructure of a 6061 aluminum alloy after equal channel angular pressing. *Scr. Mater.* **2014**, *45*, 901–907. [CrossRef]
31. Hockauf, M.; Meyer, L.W.; Zillmann, B.; Hietschold, M.; Schulze, S.; Krüger, L. Simultaneous improvement of strength and ductility of Al-Mg-Si alloys by combining equal-channel angular extrusion with subsequent high-temperature short-time aging. *Mater. Sci. Eng. A* **2009**, *503*, 167–171. [CrossRef]
32. ASTM E399-12e3. In *Standard Test Method for Linear-elastic Plane-Strain Fracture Toughness KIc of Metallic Materials*; American Society for Testing and Materials: West Conshohocken, PA, USA, 1970.
33. Hockauf, M.; Schönherr, R.; Wagner, S.; Meyer, L.W.; Podlesak, H.; Mücklich, S.; Wielage, B.; Krüger, L.; Hahn, F.; Weber, D. Equal-channel angular pressing of medium- to high-strength precipitation hardening aluminium wrought alloys. *Mat.-Wiss. Werkstofftech.* **2009**, *40*, 540–550. [CrossRef]
34. Shang, J.K.; Ritchie, R.O. On the particle-size dependence of fatigue-crack propagation thresholds in SiC-particulate-reinforced aluminium alloy composites: Role of crack closure and crack trapping. *Acta Metall.* **1989**, *37*, 2267–2278. [CrossRef]

35. Hockauf, K.; Halle, T.; Hockauf, M.; Wagner, M.F.-X.; Lampke, T. Near-threshold fatigue crack propagation in an ECAP-processed ultrafine-grained aluminium alloy. *Mater. Sci. Forum* **2011**, *667–669*, 873–878.

36. Shang, J.K.; Yu, W.; Ritchie, R.O. Role of silicon carbide particles in fatigue crack growth in SiC-particulate-reinforced aluminum alloy composites. *Mater. Sci. Eng. A* **1988**, *102*, 181–192. [CrossRef]

metals

MDPI

Article

Effect of Strain Localization on Pitting Corrosion of an AlMgSi0.5 Alloy

Daniela Nickel [†], Dagmar Dietrich [†,*], Thomas Mehner, Philipp Frint, Dagobert Spieler and Thomas Lampke

Technische Universität Chemnitz, Professur Oberflächentechnik/Funktionswerkstoffe, 09107 Chemnitz, Germany; daniela.nickel@mb.tu-chemnitz.de (D.N.); thomas.mehner@mb.tu-chemnitz.de (T.M.); philipp.frint@mb.tu-chemnitz.de (P.F.); dagobert.spieler@mb.tu-chemnitz.de (D.S.); thomas.lampke@mb.tu-chemnitz.de (T.L.)

* Author to whom correspondence should be addressed; dagmar.dietrich@mb.tu-chemnitz.de;
  Tel.: +49-371-531-35392; Fax: +49-371-531-835392.
† These authors contributed equally to this work.

Academic Editor: Heinz Werner Höppel
Received: 8 December 2014; Accepted: 29 January 2015; Published: 3 February 2015

Abstract: The corrosion susceptibility of an age-hardened aluminum alloy in different processing conditions, especially after a single pass of equal-channel angular pressing (ECAP), is examined. The main question addressed is how corrosive attack is changed by strain localization. For that purpose, an AlMgSi0.5 alloy with a strain localized microstructure containing alternating shear bands was subjected to potentiodynamic polarization on a macro-scale and micro-scale using the micro-capillary technique. Pitting potentials and the corrosion appearance (pit depth, corroded area fractions and volumes) are discussed with respect to microstructural evolution due to casting, extrusion and ECAP. Size, shape and orientation of grains, constituent particle fragmentation, cell size and microstrain were analyzed. Stable pitting of shear bands results in less positive potentials compared to adjacent microstructure. More pits emerge in the shear bands, but the pit depth is reduced significantly. This is attributed to higher microstrains influencing the stability of the passivation layer and the reduced size of grains and constituent particles. The size of the crystallographic pits is associated with the deformation-induced cell size of the aluminum alloy.

Keywords: AlMgSi0.5 alloy; ECAP; pitting corrosion; micro-capillary cell; deformation localization

---

1. Introduction

Aluminum alloys are used in a wide range of technical applications, especially when considering the lightweight aspect. Mechanical properties are influenced by microstructural heterogeneities formed by adding alloying elements mostly combined with thermal or thermo-mechanical treatments. Solution heat treatment and aging lead to the formation of solid solutions and intermetallic compounds (ICs) stimulating mechanisms of solid solution and precipitation strengthening. Mechanical-based treatments cause strain hardening by multiplication of dislocations or grain refinement in the deformed metal. An effective deformation technique is equal-channel angular pressing (ECAP), also known as equal-channel angular extrusion, originally invented by Segal in 1972, first described in his Sc.D. thesis in 1974 and patented in 1977 (Patent of the USSR, No. 575892) [1,2]. A massive metallic billet is pressed through two intersecting channels whereby the material is severely deformed by exceptional, intensive and oriented simple shear, which is localized inside thin, so-called shear bands [1]. Accordingly, after one ECAP pass, one family of shear bands is activated. Repetitive pressing induces new shear bands depending on the rotation between the successive passes. Grain refinement down to ultrafine grains (UFG), re-distribution and fragmentation of second phases or impurities if existing in the material and texture evolution differ according to the applied processing route. In addition, the bulk material is characterized by high densities of dislocations and grain boundaries, internal elastic strains as well as residual stress thus delivering improved mechanical properties [3,4].

The effect of any microstructural heterogeneity results in an electrochemical heterogeneous behavior influencing the susceptibility to corrosion. In recent years, significant progress has been achieved in the understanding of both the microstructure of high strength alloys and its influence on corrosion [5]. Dominant features like the surface fraction of grain boundaries and dislocations as well as the distribution of second phases including constituent and impurity particles, dispersoids and precipitates are studied with respect to their corrosion susceptibility. The improved physico-chemical characterization of the composition and spatial distribution of second phases together with advanced electrochemical techniques expanded the understanding of the influence of intermetallic particles on the corrosion of conventionally grained (CG) alloys. A comprehensive review is given by Hughes *et al.* [5]. Recently, electrochemical properties of UFG aluminum and its alloys in different ECAP conditions were studied in a variety of publications (Table 1). Pitting corrosion starts with the local destruction of the natural oxide layer by critical anions (e.g., chloride ions) and is a dominant corrosion type of passivating materials. In electrochemical studies used to investigate the susceptibility to pitting, the pitting potential marks the breakthrough of the passive layer. In high-purity aluminum and single-phase aluminum model alloys, the pitting potential is lowered [6,7] or unaffected [8] compared to the CG material and is ascribed to the effect of grain refinement. Ralston *et al.* [9] correlate the corrosion rate in terms of oxide formation and film ion conduction with reactive surfaces provided by enhanced grain boundary density. Brunner *et al.* [7] found a lowered pitting potential due to the transition from CG to UFG material but no significant influence with the number of ECAP passes. The pit morphology after polarization occurred less crystallographically, deeper and more localized in the UFG material compared to the CG material in which laterally spreading crystallographic filiform corrosion dominated. The dislocation density and the deformation state were supposed to be the dominating factors for the different propagation mode of pitting corrosion: propagation in depth for UFG and laterally spreading for CG material [7]. The variation in passive film stability and the existence of local galvanic cells stimulate the anodic dissolution of the surrounding matrix material. Coherent solute enriched clusters can act as local cathodes in precipitation hardening alloys. Ralston *et al.* [10] found a critical particle size (of some nm) below which they do not act as distinct galvanic cells, but promote susceptibility to pitting if they increase.

Table 1. Review of pitting corrosion results published for Al and its alloys.

Material and processing	Effect of pass number on E_{pit}	Dominant pit initiation	Pit morphology with increasing pass number	Analytical techniques	Reference
high purity Al 2, 6, 10 ECAP passes (route Bc)	lowered	dislocations, grain boundaries		PP	[6]
Al-2.5 wt% Cu-1.5 wt% Mg alloy aging, no ECAP	unaffected/lowered	critical size of GP zones (solute enriched regions)		PP, CT	[10]
Al-Mg model alloys (varying Mg content) 4, 8, 12 ECAP passes (route Bc)	lowered/unaffected	deformation state more dominant than grain size	after PP: pits deeper and more localized, hindered repassivation with increasing pass number, filiform, decreasing crystallographic character	PP, EIS	[7]
high purity Al 1, 4, 8 ECAP passes (route Bc)	unaffected	more reactive surfaces for passivation (grain size, grain boundaries)		PP	[8]
AA 1050 1, 2, 3, 5 passes	raised	fragmentation of Si-containing constituent particles	after PT: decreased pit size, less severe attack	PP, PT, EIS	[11]
AA 1100 8 ECAP passes (route Bc); AA5052 4 ECAP passes (route Bc)	raised	fragmentation of Fe-Al impurities, increased formation rate of passivation film	after PT: decreased pit size, crystallographic pits	PP, EIS	[12]
Al-5.4 wt% Ni and Al-5 wt% Cu 2, 4, 6 ECAP passes (route Bc)	raised	fragmentation of constituent particles, microstructural homogenization	after PP: fewer pits in the shear bands	PP	[6]
EN AW-6082 1, 2, 8 ECAP passes (route E), 1 ECAP pass + post ageing	raised	fragmentation and redistribution of constituent particles	after PP: decreased pit depth	PP, EIS, CV	[13,14]
low purity Al (Al-0.5 wt% Si-0.15 wt% Fe) 16 ECAP passes	raised	more stable passivation film due to strain induced crystalline defects and residual (micro)stress	after PT: decreased pit size uniform distribution	PP, PT, IT, EIS	[15]
low purity Al (99.1 wt%) with α-AlFeSi phases 2, 4 ECAP passes (route A)	lowered	fragmentation of constituent particles (α-phase), grain refinement and higher dislocation density	after IT: crystallographic pitting	IT, PP, EIS	[16]
low purity Al (99.1 wt%) with Al8Fe2Si phase 1, 2, 3, 4, 5 ECAP passes (route A)	unaffected	fragmentation of constituent particles, increased volume fraction of grain boundaries	after IT and PP: reduced intergranular corrosion increased pitting increased density and decreased pit depths	IT, PP, EIS, 3D optical profilometry	[17]

Note: PP, potentiodynamic polarization test; PT, potentiostatic polarization test; EIS, electrochemical impedance spectroscopy; CV, cyclovoltammetry; CT, current transient experiments; IT, immersion test; Epit, pitting potential.

Constituent particles in multi-phase materials mostly act as cathodic inhomogeneity performing a key role in pit initiation and propagation. With an increasing number of ECAP passes, the pitting potential of a variety of multi-phase aluminum alloys, which were reported to increase [11,12,15], decrease [16] or stay unaffected [17]. Correspondingly, for AA 5052, after four ECAP passes, the pitting potential was distinctly shifted to more positive potentials [12]. The crystallographic pits were smaller and more uniformly distributed in the UFG material than in the as-cast condition. The authors suggested a higher oxide formation rate based on the enhanced grain boundary fraction and dislocation density in the UFG materials. Similarly, Song *et al.* [15] stated that strain-induced crystalline defects provide more nucleation sites for the formation of a denser and thicker oxide film. Akiyama *et al.* [6] found an increasing pitting potential of Ni and Cu containing aluminum alloys due to ECAP processing. Fewer pits in the shear band areas were attributed to the microstructural homogenization and refinement of primary particles [6]. The amount, fragmentation and redistribution of constituent particles, containing mostly Si and Fe along with Al, were considered as the main factors for the modified pitting susceptibility and morphology of multiphase alloys [6,11,12,16].

Korchef *et al.* [16] found decreasing pitting potentials explained by the increasing fragmentation of α-AlFeSi particles acting as cathodes and driving anodic dissolution of the surrounding matrix. Jilani *et al.* [17] found no influence on the pitting potential with an increasing number of ECAP passes, but an increasing pitting density and decreasing pit depth. The change in the pitting morphology was assumed because of the fragmentation of Al_8Fe_2Si particles and/or the related increase of the volume fraction of grain boundaries acting as potential sites for pitting. Finally, the corrosion resistance of UFG materials processed by accumulative roll bonding shows similarly contrasting results [18,19].

Table 1 presents the reviewed literature together with own results [13,14] demonstrating little consensus in the influence of ECAP processing on the corrosion behavior of aluminum alloys even within the same alloy. Certainly, the main challenge of corrosion studies in this regard is the lack of a standardized material condition suitable as benchmark for evaluating the effect of ECAP processing on the corrosion behavior.

In order to overcome this obstacle, the present work adapts the testing and evaluating area to the strain localizations generated by a single ECAP pass processing. The selected ECAP condition presents concurrently ECAP-affected (shear bands) and non-affected (matrix) areas, which can be locally tested using a micro-capillary cell technique [20] and independently evaluated because only one family of shear bands is activated. Additional impacts by further processing are excluded. A commercial alloy AlMgSi0.5 with a well-known microstructure [21–23] was selected. The results of the micro-scale electrochemical studies are compared to standard macro-scale experiments including the material conditions of the foregoing processing steps casting and extrusion. Finally, the induced pitting damage is separately quantified for shear bands and matrix zones. The results are correlated with the microstructure of the material.

2. Experimental Section

2.1. Sample Preparation

A commercial AlMgSi0.5 alloy (EN AW-6060, chemical composition given in Table 2) was used in different material conditions. Samples were taken from a cast billet (referred to as cast) and from bars after appropriate processing steps. Homogenization and solution annealing at 560 °C for 10 h was followed by water quenching. The ingot was extruded at room temperature (RT) reducing the initial diameter of 110 mm to a cross-section of 50×50 mm^2 (samples referred to as RT-extruded). Severe plastic deformation was achieved using the combination of RT-extrusion with a subsequent large-scale ECAP single pass at RT [22,23] (samples referred to as ECAP). The internal angle of the die was 90°, the pressing speed 20 mm min^{-1} and the backpressure 200 MPa. The resulting effective strain was approximately 1.1 [24].

Table 2. Chemical composition of the investigated EN AW-6060 (wt%).

	Si	Fe	Cu	Mn	Mg	Zn	Al
EN AW-6060	0.45	0.16	0.003	0.026	0.49	0.003	bal.

Samples with longitudinal planes were cut of each billet. After mechanical grinding with SiC, the final polishing step of the metallographic preparation was dependent on the intended investigation method. For optical microscopy (OM), polishing in a silicon oxide suspension (OP-S, Struers GmbH, Willich, Germany) and grain boundary etching for microstructural contrasting (2% NaOH) was applied. For electrochemical measurements, polishing with 1 μm diamond suspension was used to achieve a fairly chemically unaffected surface. After corrosion experiments, the surface was carefully rinsed with de-ionized water and ethanol and subjected to vibrational polishing using OP-S to provide a clean and deformation-free surface for backscattered electron (BSE, Zeiss, Jena, Germany) imaging and electron backscatter diffraction (EBSD, AMETEK GmbH, Wiesbaden, Germanyin the scanning electron microscope (SEM, Zeiss, Jena, Germany [25].

2.2. Electrochemical Measurements

Deformation localizations (shear bands) were detected by OM on the etched ECAP sample surface and marked by indentations. The electrochemical experiments were accomplished in an aerated and unbuffered 0.1 M NaCl solution (pH 7) at (22.0 ± 0.5) °C. The samples were exposed to the solution for 15 min to stabilize the open circuit potential (OCP). Polarization curves were measured using a high-resolution potentiostat IMP83 PCT-BC (Jaissle Elektronik, Waiblingen, Germany). After determining the starting potential, potentiodynamic polarization experiments were done covering the potential range from 200 mV below the corrosion potential (between −1300 mV and −1100 mV depending on the material condition) to 1000 mV with a scan rate of 1 mVs^{-1}. For experiments with an exposed area of 10 mm in diameter (referred to as macro-scale), a three-electrode corrosion cell with a platinum plate as counter electrode and an Ag/AgCl (3 M KCl) reference electrode was used according to the standard specification DIN 50918. Macro-scale measurements were repeated three times.

For local micro-electrochemical measurements (referred to as micro-scale), a glass capillary with a ground tip of 200 µm in diameter was used. A layer of silicone rubber worked as a seal to prevent leakage. The capillary size was selected in order to cover either the shear band or the matrix of the material in ECAP condition. The three-electrode set-up consisting of the glass capillary, a 0.5 mm platinum wire counter electrode and a Ag/AgCl (3 M KCl) reference electrode was mounted in an objective nosepiece. With the samples fixed on the stage of the OM, exact positioning on the microstructural localization (shear band or matrix) was obtained. The set-up of the micro-cell system is described in detail in [20,26]. Micro-scale measurements were repeated nine up to 13 times depending on the material condition.

2.3. Microstructural Characterization

Characterization of the microstructure and the pit morphology were performed in a field-emission scanning electron microscope (FE-SEM, NEON40EsB, Carl Zeiss MicroImaging GmbH, Jena, Germany) equipped with an EBSD system (EDAX TSL OIM 5.2, AMETEK GmbH, Wiesbaden, Germany. EBSD patterns were acquired at an acceleration voltage of 15 kV with a 60 µm aperture in high-current mode under a sample tilt angle of 70°. After data acquisition, a slight cleanup procedure for the nearest neighbor pixels was conducted comprising neighbor confidence index (CI) correlation and grain CI standardization with a minimum CI of 0.1.

Material removal by corrosion was quantitatively evaluated by using a 3D optical profilometer (Mikro-CAD compact, GFMesstechnik GmbH, Teltow, Germany). Corroded surface areas were determined by means of the software a4i analysis (GFMesstechnik GmbH, Teltow, Germanycity). Depth and volume of the resulting corrosion pits were measured using replicas of the tested regions. Pits with a depth larger than 5 µm were considered.

X-ray diffraction (XRD) analysis (sin2ψ method) was applied for the determination of local residual stresses and microstrains. The diffractometer D8 Discover (Bruker AXS, Karlsruhe, Germany) was equipped with a Cu anode, a Ni filter for monochromatization, polycap optics and a 2D detector (Vantec-500, Karlsruhe, Germany. The {311} lattice planes of Al were considered for the residual stress measurements using several measurement directions (0°, 45°, 90°–rotation around the sample normal) each with tilt angles between −50° and 50° (in 10° steps) and a measurement time of 900 s for each angle. A small aperture (100 µm) in the primary beam, adequate to the dimensions of the strain localizations, and a point detector (SOL-XE) with a 0.6 mm detector slit were used for the determination of microstrains and coherence lengths in a line scan. The step size of the line scan was 100 µm. The program LEPTOS was used for stress evaluation. The line widths of the strong textured {220} lattice planes were evaluated by means of the program TOPAS under consideration of the device-related line-broadening effects to estimate the coherence length.

3. Results and Discussion

3.1. Evolution of Microstructure

The cast material (Figure 1a) is characterized by equiaxed grains with a mean diameter of 100 μm and numerous constituent particles decorating the grain boundaries. The IC particles in the cast condition are a few micrometers in size; homogenization and solution annealing resulted in a slight enlargement. The extrusion process transformed the alloy microstructure to elongated grains with an aspect ratio around 0.5 associated with the redistribution of the IC particles located at the grain boundaries. The single-pass ECAP process introduced a series of deformation localizations (shear bands) intercalated in fairly unchanged matrix regions. The grains in the shear bands become more elongated to an aspect ratio below 0.3. The IC particles are re-aligned with the aluminium grains and slightly fragmented (Figure 1b). Additionally, the deformation processing introduced a sub-grain structure with a size around 1 μm (Figure 2c,d). This is in agreement with transmission electron microscopic results achieved at the same alloy [23].

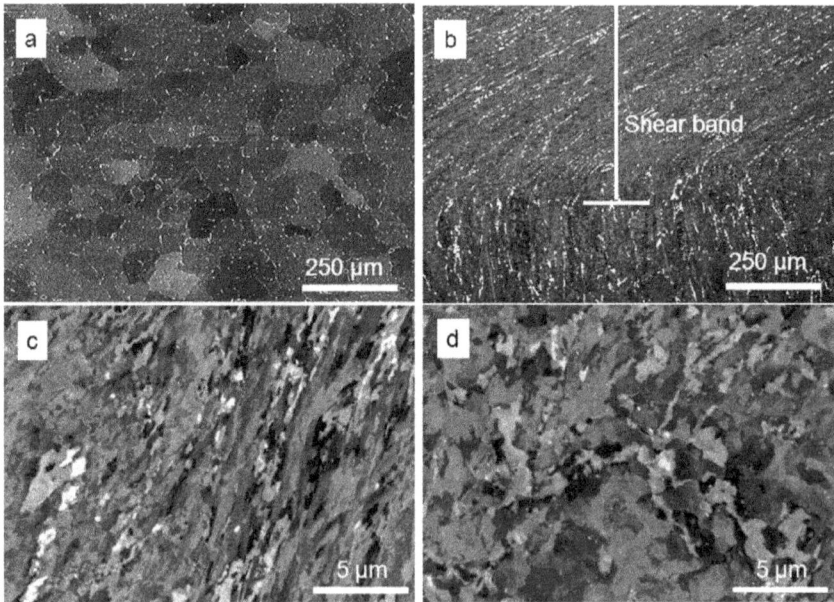

Figure 1. Microstructure of the AlMgSi0.5 alloy in (**a**) cast condition showing constituent particles in the grain boundaries; (**b**) ECAP (equal-channel angular pressing) condition showing shear bands (a part of it is marked by a white bar) introduced into the matrix (below); (**c**) elongated grains in the shear band and (**d**) fairly unchanged grains in the matrix, which is similar to the extruded condition.

According to Birbilis *et al.* [27], the IC particles of particular interest with regard to localized corrosion are those with the greatest quantity in size or frequency—these are AlFeSi phases in the AlMgSi0.5 alloy. As shown for the extruded condition of the alloy [28], the iron and silicon containing constituent particles include compositions with different stoichiometry. Apart from the AlFeSi phases, the solid solution matrix of the as-cast billet contains dissolved Mg and Si, which precipitate to fine β and β′ particles during quenching [29,30]. Certain observations show that the hardening precipitates in the studied AlMgSi0.5 alloy undergo fragmentation and change in coherency by ECAP [31,32].

Apart from grain sizes and aspect ratios, orientation maps (Figure 2a–c) were derived from EBSD measurements which illustrate the microstructural evolution of the material with respect to grain orientation or texture. The random orientation of the cast material (Figure 2a) is transformed by RT extrusion to a fiber texture with the main component <110> in extrusion direction (Figure 2b). The grain orientation map of the ECAP condition (Figure 2c) shows a shear band in the center with a texture fairly similar to the matrix. Deformation localizations are clearly reflected by dark regions in EBSD pattern quality maps as shown later in Figure 9.

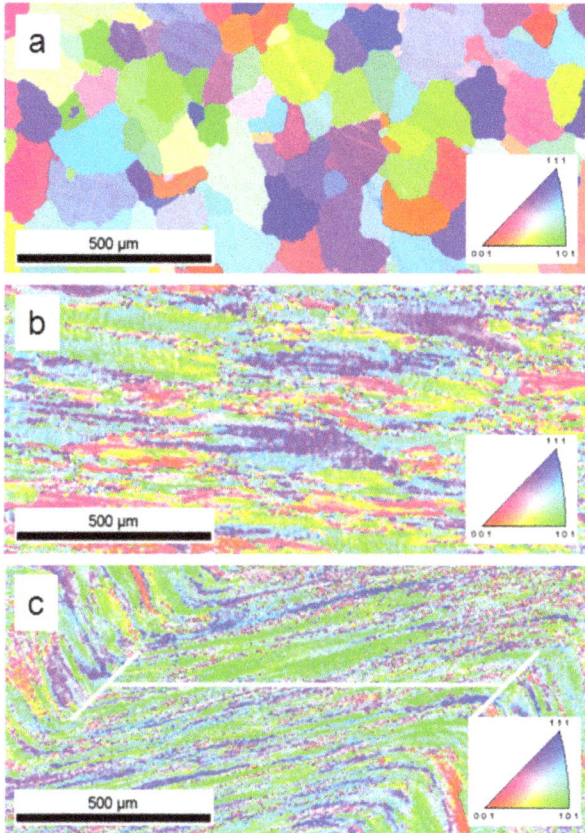

Figure 2. Orientation distribution maps of the material in the (**a**) cast; (**b**) the RT-extruded and (**c**) the ECAP condition showing the grain shape transformation to elongated grains of reduced size and texture evolution by extrusion and ECAP; the shear band in (**c**) is marked by a white bar.

The measurement of residual stresses and microstrains was focused on the deformation localizations of the material in ECAP condition. Almost no differences could be detected for the principal stress components σ1 and σ2 inside shear bands and matrix regions. The measured components of the residual stress are σ1 ≈ 60 MPa and σ2 ≈ −20 MPa.

Microstrains have been specified referring to the full width at half maximum of the Gaussian part of the line profile. Measurements of several points on the material in the RT-extruded condition were taken for comparison and reveal a fairly constant microstrain of 0.21. The microstrain in the ECAP condition varies between 0 and 0.34. A line scan was measured with an increment of 100 µm across three shear bands intercalated in the matrix. Figure 3 shows the local distribution of microstrain in correlation to the shear bands revealed by grain boundary etching. The shear band-matrix transitions are marked by white lines. It can be seen that the widths of deformation localizations comprise a wide range. For reasons of representation, the values of the microstrain are grouped in bins. Color-coded according to the legend, the microstrain scan is inserted in the optical micrograph at the position of the measurement. As expected, higher values of microstrain are found in the shear bands and lower values in the matrix (Figure 3). The broad matrix band on the right shows an additional subdivision—possibly because of the actual three-dimensional distribution of shear bands, which is not observable in the plain cross-section. Apparently, the microstrains are a result of the deformation localization introduced by the severe deformation during the ECAP pass.

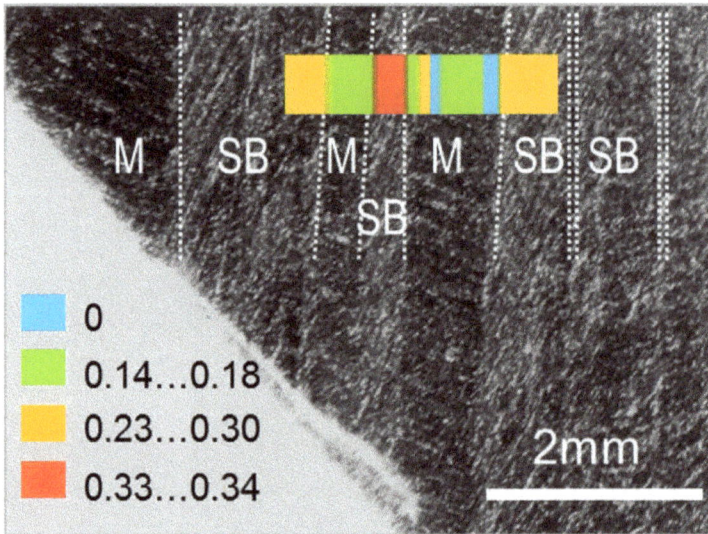

Figure 3. Local distribution of the microstrain in a line scan across three shear bands of the material in the ECAP condition; the region of interest is shown with respect to the deformation localizations shown by grain boundary etching (for color-coding, see legend).

Crystallite size was determined by the Lorentzian part of the line profile. The measured characteristic is the distance that scatters the incident X-ray beam coherently. Apart from grains and sub-grains defined by high- and low-angle boundaries, coherent regions are bounded by accumulated dislocations and are commonly assigned as cells. Cell sizes are averaged out at 180 nm in the extruded condition and typically range from 80 nm–170 nm for both matrix and shear bands of the ECAP condition, which is consistent with the published cell size data determined by STEM [32]. Interestingly, the mean cell size in the matrix region is smaller ($\approx$100 nm) compared to the shear band with the largest cells in the region with the highest microstrain.

3.2. Pitting Corrosion Potential Derived from Macro- and Micro-Electrochemical Tests

Pitting corrosion is the most common corrosion type for aluminum alloys. The local destruction of the passivation film is caused by anions, e.g., chloride ions. Potentiodynamic polarization curves were measured to evaluate pitting susceptibility. The most negative potential above which pits nucleate and grow (pitting potential) was determined as the decisive corrosion characteristic. Figure 4 gives an example for the curves measured on the material in ECAP condition. A grouping of the pitting potentials depending on the position of the micro-capillary either on the shear bands or on the matrix is observed.

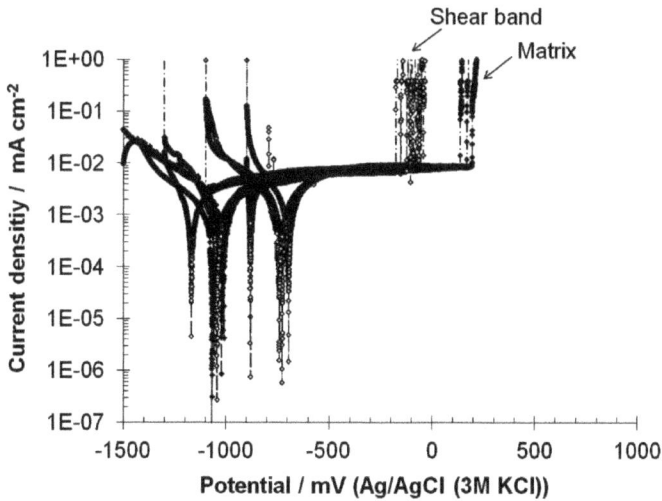

Figure 4. Potentiodynamic polarization curves measured on the shear bands and the matrix.

Pitting potentials and the corresponding corrosion appearance are shown in relation to the material condition and to the exposed area of 10 mm (macro-scale) and 200 μm (micro-scale) in diameter in Figure 5. The pitting potentials determined by macro-scale electrochemical measurements are very similar regardless of the material condition (Figure 5). Averaged out −430 mV referring to Ag/AgCl (3 M KCl), a variation of ±30 mV for the cast condition, of ±90 mV for the RT-extruded condition and of ±40 mV for the ECAP condition is ascertained. This is in agreement with the results of [8] and [17], which stated that the pitting potential of low purity aluminum does not vary with grain size or with the size of the Al_8Fe_2Si phase. The macro-scale pitting potentials are shifted to at least 700 mV lower values compared to micro-scale pitting potentials. This potential shift was already evidenced for a 2024 alloy in T3 temper state [33] and for Al-Zn-Mg-Cu alloys [34]. Evidently, this is because of the number of pit initiating constituent particles present in the testing area. More constituent particles exist in the area exposed in macro-scale experiments compared to the lower number in the area exposed in micro-scale experiments. For the same reason, the pitting potentials of the micro-scale measurements are spread over a wider potential range, also mentioned in [27].

Micro-scale experiments result in pitting potentials from −80 mV to +340 mV for the cast condition, more negative potentials from −160 mV to +300 mV for the RT extrusion condition and similarly from −180 mV to +270 mV for the ECAP condition. The scatter of the pitting potentials observed for all material conditions is suspected to be mainly the effect of various surface fractions of second phases in the exposed area according to Birbilis *et al.* [27]. Furthermore, the micro-scale experiments show a clear grouping of the higher pitting potentials related to the matrix (between +100 mV and +270 mV) and lower pitting potentials related to the shear bands (between −180 mV and +50 mV), whereas no influence of material conditions can be seen in the macro-scale experiments. Thus, micro-scale experiments allow for the differentiation of the pitting potentials in ECAP induced deformation localizations. The selected shear bands exceed a width of 600 μm thus providing a test region surely larger than the used opening of the micro-capillary (200 μm in diameter). Compared to the matrix, the significantly lower pitting potential in the shear band indicates more weak points with respect to corrosion in this region.

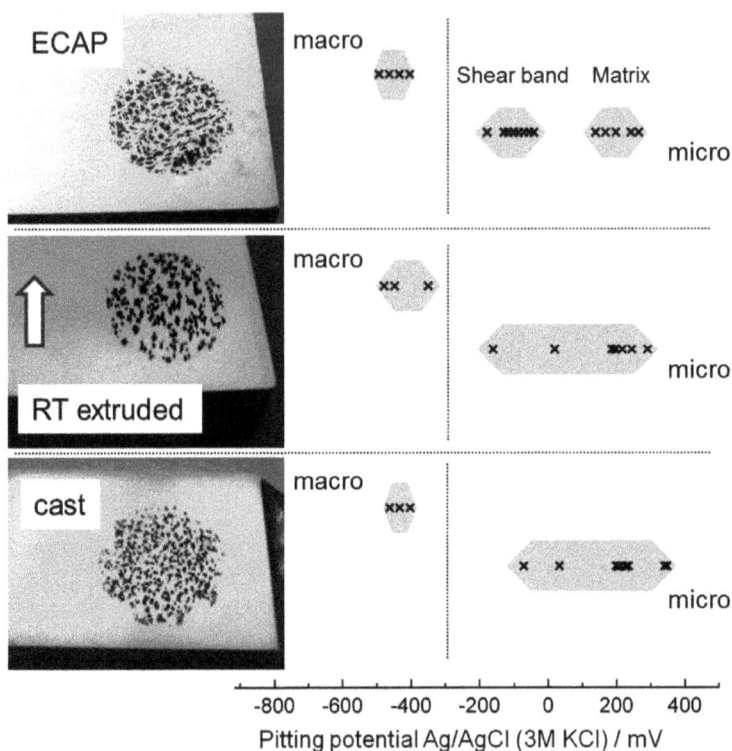

Figure 5. Documentation of the corrosion appearance after macro-scale potentiodynamic polarization of the ECAP condition, the extruded condition (extrusion direction marked by arrow) and the cast condition showing the modification from uniform to oriented (left from top to bottom) with corresponding pitting potentials in 0.1 M NaCl determined by macro-scale tests (middle) and micro-scale tests (right) at 1 mA cm^{-2}.

3.3. Pit Morphology

In Figure 6a–c, the secondary electron images (top) and the corresponding orientation maps derived from EBSD (below) show typical examples of the corrosion appearance as a consequence of different processing history. Pitting seems to start as well as to be terminated at the precipitation decorated grain boundaries thus evolving different corrosion appearance with respect to the grain shape.

The dominant feature of the alloy microstructure along with the modification of grain size and grain shape is the distribution and fragmentation of second-phase intermetallic particles (constituent particles) with respect to electrochemical characteristics. The constituent particles in an AlMgSi alloy are electrochemically nobler compared to the surrounding aluminum solid solution [16,17,27]. Due to the galvanic coupling between the AlSiFe phases and the surrounding Al matrix in the Cl$^-$-containing test environment, the naturally occurring passivation layer will be reduced and the surrounding aluminum dissolved. Starting at the precipitations mostly localized at the grain boundaries, the corrosion paths regularly start at the grain boundaries, spread into the grain and are terminated at the opposite grain boundary when the constituent particles are undermined and lose electrical contact. The large equiaxed grains of the cast condition are mirrored by a similar corrosion appearance. Similarly, the aligned and elongated grains of the material in the RT extrusion and ECAP condition are associated with an aligned and elongated corrosion appearance. The fragmentation of constituent particles by ECAP processing provides numerous starting sites for corrosion, but also faster removal of the smaller particles by undermining the surrounding aluminum. Thus, the propagation of the corrosion attack in vertical direction is decreased.

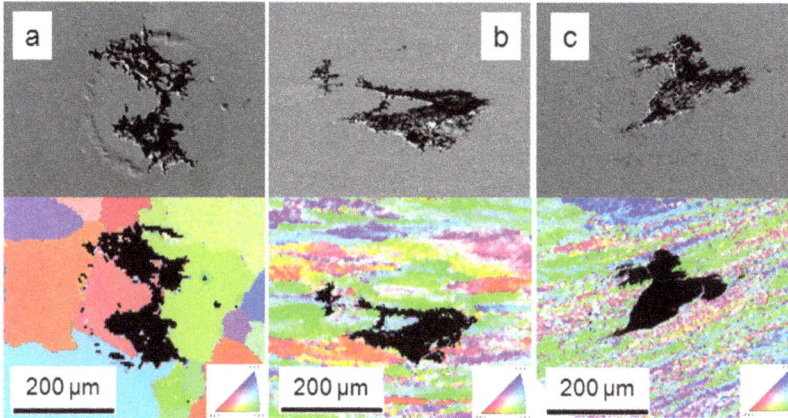

Figure 6. Corrosion appearance after micro-scale potentiodynamic polarization of the (**a**) cast condition; (**b**) extruded condition (extrusion direction is horizontal) and (**c**) in a shear band of the ECAP condition.

The pitting appearance produced during macro-scale testing shows a striking correlation to the microstructure developed by the processing history (Figure 5). The uniform corrosion in the cast condition is transformed to a non-uniform appearance corresponding to the directed and elongated grains after extrusion and ECAP. This behavior is accompanied by an obviously shallower corrosion attack in the ECAP condition which will be quantitatively evaluated and discussed later on. Crystallographic pitting is evidenced regardless of the material condition but with a remarkable influence on the size of the individual {100} facetted pits. Large crystallographic pits around 1 µm in length of the edge can be seen in the cast condition (Figure 7a), around half as long are the edges of crystallographic pits in the in RT-extruded condition (Figure 7b) and even more reduced in the ECAP condition (Figure 7c,d). Particularly noteworthy is the fact that the reduced size of individual crystallographic pits in the matrix zones in the ECAP condition is more similar to those in the shear bands. This was unexpected, since the mean grain size and the primary precipitate size are fairly equal in the RT-extruded condition and in the matrix of the single pass ECAP condition.

ECAP processing enhances the fraction of low- and high-angle grain boundaries, the dislocation density and the fragmentation of strengthening precipitates of the studied aluminum alloy [23,32]. A bimodal grain size distribution and an increased dislocation density appear already after one ECAP pass in both matrix and shear bands [32]. The dislocations accumulate to a complex network of deformation induced cell boundaries [23]. A schematic representation of the cell distribution and corresponding STEM images are shown in Figure 8 (adopted from [23]). The size of the cells as the smallest elements which subdivide grains is evaluated by XRD in this work and ranges between 80 nm and 180 nm.

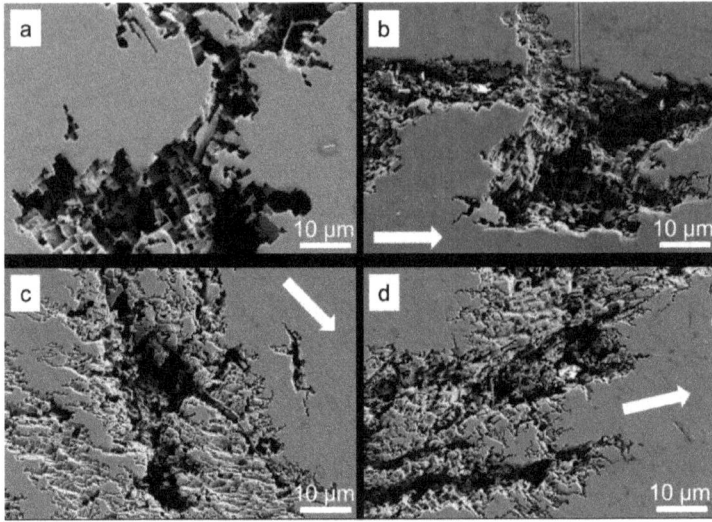

Figure 7. Formation of crystallographic {100} facetted pits due to micro-scale potentiodynamic polarization in the (**a**) cast condition; (**b**) in the extruded condition (extrusion direction and grain alignment marked by arrow); (**c**) in ECAP condition in the matrix and (**d**) in a shear band (deformation direction marked by arrows).

Apart from the decreasing pitting potential in shear bands, we observe a significant reduction of the single crystallographic pit size in both the matrix and shear bands in the ECAP condition. This observation correlates with the reduced dimensions of the cells evaluated by XRD and STEM. We conclude that the cell size and the microstrains are the distinguishing features to control the pit growth in the ECAP condition.

The reduced corrosion depth after ECAP was already indicated by several authors [11–15,17]. Contrary results were reported for Al-Ni and Al-Cu alloys containing other second phases [6]. In an Al-Mg model alloy [7], more localized and deeper pits were found in the UFG condition. Furthermore, the pitting potential evidenced influence of the number of ECAP passes. However, tests after one ECAP pass were not considered.

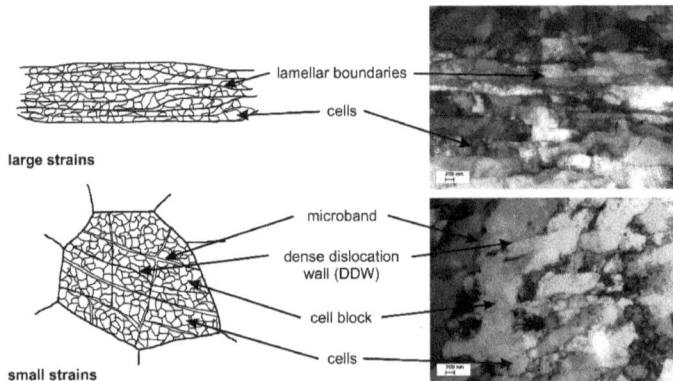

Figure 8. Schematic representation of the deformation microstructure and grain subdivision at different strains (large/small) and corresponding STEM-micrographs of the investigated aluminum alloy [23].

Jilani *et al.* [17] qualitatively discussed the change in the corrosion appearance from a low density of deep pits to a high density of less deep pits with an increasing ECAP pass number. This is in agreement with our observation. At first sight, the removal of material in the ECAP condition seems to be more concentrated in the regions without deformation localizations. The simple comparison of corrosion areas leads to this conclusion, but the volume of the eroded material has to be considered. To substantiate our findings by quantifying the material removal, several metrological characteristics were determined using a 3D optical profilometer. A depth mapping of the complete area exposed to macro-scale testing after ECAP is shown in Figure 9 with a magnified detail in the inset. Pit depths between 5 µm (red) and 75 µm (blue) were recorded and are shown by a color code. The contiguous EBSD pattern quality map corresponds to the position of the detailed depth map and illustrates the deformation localizations by darker gray tones (the corroded areas are shown in black). Several metric features were analyzed to reveal differences in the matrix and shear bands and to compare the other material conditions: the fraction of the corroded area, the corroded volume, the average corrosion depth, the maximal pit depth, the average pit depth and the density of pits (Table 3). The average corrosion damage refers to a uniform removal of the material across the tested area. The average pit depth refers to the total corroded volume divided by the total corroded area. The pit density is related to the total area used for macro-scale testing.

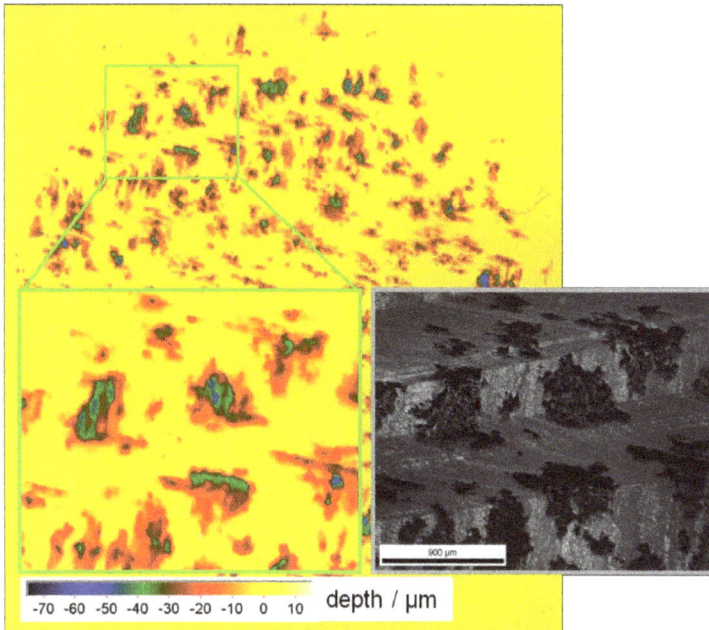

Figure 9. Depth mapping of pitting corrosion after macro-scale potentiodynamic polarization of the ECAP material (depth scale color-coded); the left inset shows the details; the right inset shows the corresponding EBSD pattern quality map indicating deformation localization (position of the insets is marked by the box in the overview map).

Table 3. Characteristics of the corrosion appearance for the studied conditions after macro-scale potentiodynamic polarization (Ø 10 mm).

State	Corroded area fraction (%)	Corroded volume (mm³)	Average corrosion depth (μm)	Maximal pit depth (μm)	Average pit depth (μm)	Density of pits (pits/mm²)
cast	34	0.57	7.3	100	19.6	4.6
RT-extruded	37	0.75	9.6	120	28.1	3.9
ECAP						
sum	38	0.43	5.1	75	13.5	8.7
matrix	45	0.24	6.0	75	13.3	8.0
shear band	32	0.19	4.4	75	13.6	9.3

The fraction of the corroded surface area of the cast condition is about 10% smaller than that of the RT-extruded and ECAP condition, but the corroded volume follow the order ECAP < cast < RT-extruded. The same applies to the average corrosion depth, the maximal pit depth and the average pit depth. Hence, the corrosion of the RT-extruded condition is the most local one of the considered conditions and thus most susceptible to corrosion damage. In contrast, the material in the ECAP condition corrodes in a more uniform way with remarkably lower maximal and average pit depths compared to the other conditions. The pit densities verify the statement of the more uniform corrosion of material in the ECAP condition. *Vice versa*, lower pit densities in association with enhanced corroded volumes for the cast and the RT-extruded conditions correspond to more local corrosion.

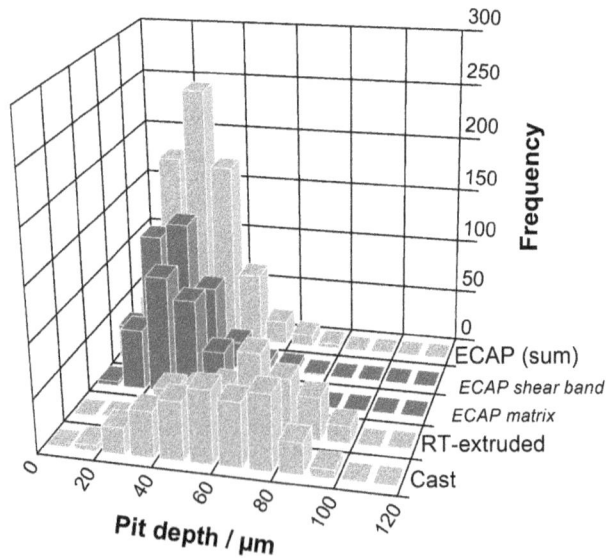

Figure 10. Frequency distribution of pit depths for different material conditions, additionally matrix and shear band are considered separately.

In order to substantiate the result, pit depth frequencies are shown for the cast, RT-extruded and ECAP condition with a detailed breakdown of the matrix and shear band (Figure 10). The pit depths are grouped in 10 μm bins. A fairly similar broad distribution of pit depths between 10 μm and 100 μm was found for the cast and the RT-extruded conditions. The maximum of the frequency distribution is about 65 μm. A tendency to lower pit depths in the cast condition can be noticed. The ECAP condition shows a strikingly enhanced number of pits. More than twice as many pits are evidenced in the material after one ECAP pass, but the depth distribution is quite sharp with its maximum being only about 30 μm. In order to differentiate the corrosion attack between the shear bands and the matrix,

Metals **2015**, *5*, 172–191

distinct frequency distributions of pit depths are shown (Figure 10). A slightly higher number of pits in the shear band contribute to the total pit frequency distribution. However, the pit depths in the shear band tend to be lower compared to the matrix which is attributable to the fragmentation and re-distribution of the constituent particles and moreover, to the enhanced microstrains in the shear bands. Thus, offering additional sites for the pit initiation and growth, pitting corrosion results in a uniform attack and prevents local deep damage.

4. Conclusions

With the focus on deformation localizations, our study demonstrates the effect of extrusion and ECAP on pit initiation and pit growth in an aluminum alloy EN AW-6060. The refined microstructure together with a heterogeneous distribution of microstrains and an increased dislocation density are the controlling microstructural features.

The influence of deformation localizations on pit initiation is proven by local electrochemical measurements restricted to single shear bands of the alloy in the single pass ECAP condition. Deformation localizations show an enhanced pitting susceptibility in relation to the cast and extruded condition. There was a strong indication that the enhanced pitting susceptibility is related to local microstrains.

However, easier pit initiation is associated with more homogeneous and shallow pit propagation in the AlMgSi0.5 alloy after a single ECAP pass. Regardless of the material condition, dissolution of complete grains was observed starting and stopping at primary precipitates predominantly located at the grain boundaries. The origin of corrosion is the cathodic behavior of the primary precipitates with respect to the AlMgSi0.5 alloy. When the dissolution front approaches the next grain boundary, the decorating precipitates are undermined and the local electrochemical cell disappears. Hence, grain fining and precipitate fragmentation increase pitting density but reduce pitting depth.

The pitting morphology shows characteristic channels formed by {100} facetted crystallographic pits. In contrast to the uniform microstrain in the material in the extruded condition, significant differences in microstrains in the shear bands and the alloy matrix are shown and supposed to be the reason for the particular corrosion appearance. The crystallographic pits appear modified in size with regard to the material condition. The reduced size of the pits in the material in ECAP condition is attributed to the cells as the smallest elements subdividing the grains by accumulation of dislocations (cell boundaries) as a response to the microstrains generated by the single ECAP step. The cell size corresponds with the size of the crystallographic pits.

Acknowledgments: Financial support was granted by the Deutsche Forschungsgemeinschaft (German Research Foundation) via the collaborative research centre "High strength aluminium-based lightweight materials for safety components" (DFG SFB 692) as well as via the project LA 1274/27-1. The technical assistance of A. Schulze, E. Benedix, C. Gläser, K. Muhr, G. Röllig, and A. Rauscher is gratefully acknowledged.

Author Contributions: D.N. is the primary author of the paper, and performed analysis of the experimental data. D.D., T.M., P.F. and D.S. contributed to the experimental research work (SEM and EBSD, XRD, electrochemical measurements and analysis). T.L. supervised the work and discussed the results and analysis with the other authors.

Conflicts of Interest: The authors declare no conflict of interest.

References

1. Segal, V.M. Equal channel angular extrusion: from macromechanics to structure formation. *Mater. Sci. Eng. A* **1999**, *271*, 322–333. [CrossRef]
2. Segal, V.M. The Method of Material Preparation for Subsequent Working. Patent of the USSR 1977 No. 575892.
3. Vinogradov, A.; Mimaki, T.; Hashimoto, S.; Valiev, R. On the corrosion behaviour of ultra-fine grain copper. *Scripta Mater.* **1999**, *41*, 319–326. [CrossRef]
4. Valiev, R.Z.; Langdon, T.G. Principles of equal-channel angular pressing as a processing tool for grain refinement. *Prog. Mater. Sci.* **2006**, *51*, 881–981. [CrossRef]

5. Hughes, A.E.; Birbilis, N.; Mol, J.M.C.; Garcia, S.J.; Zhou, X.; Thompson, G.E. High Strength Al-Alloys: Microstructure, Corrosion and Principles of Protection. In *Recent Trends in Processing and Degradation of Aluminium Alloys*; InTech: Rijeka, Croatia, 2011.

6. Akiyama, E.; Zhang, Z.; Watanabe, Y.; Tsuzaki, K. Effects of severe plastic deformation on the corrosion behavior of aluminum alloys. *J. Solid State Electrochem.* **2009**, *13*, 277–282. [CrossRef]

7. Brunner, J.G.; May, J.; Höppel, H.W.; Göken, M.; Virtanen, S. Localized corrosion of ultrafine-grained Al–Mg model alloys. *Electrochim. Acta* **2010**, *55*, 1966–1970. [CrossRef]

8. Ralston, K.D.; Fabianic, D.; Birbilis, N. Effect of grain size on corrosion of high purity aluminium. *Electrochim. Acta* **2011**, *56*, 1729–1736. [CrossRef]

9. Ralston, K.D.; Birbilis, N. Effect of grain size on corrosion: A review. *Corrosion* **2010**. [CrossRef]

10. Ralston, K.D.; Birbilis, N.; Weyand, M.; Hutchinson, C.R. The effect of precipitate size on the yield strength-pitting corrosion correlation in Al-Cu-Mg alloys. *Acta Mater.* **2010**, *58*, 5941–5948. [CrossRef]

11. Chung, M.K.; Choi, Y.S.; Kim, J.G.; Kim, Y.M.; Lee, J.C. Effect of the number of ECAP pass time on the electrochemical properties of 1050 Al alloys. *Mater. Sci. Eng. A* **2004**, *366*, 282–291. [CrossRef]

12. Son, I.J.; Nakano, H.; Oue, S.; Kobayashi, S.; Fukushima, H.; Horita, Z. Pitting corrosion resistance of ultrafine-grained aluminum processed by severe plastic deformation. *Mater. Trans.* **2006**, *47*, 1163–1169. [CrossRef]

13. Hockauf, M.; Meyer, L.W.; Nickel, D.; Alisch, G.; Lampke, T.; Wielage, B.; Krüger, L. Mechanical properties and corrosion behaviour of ultrafine-grained AA6082 produced by equal channel angular pressing. *J. Mater. Sci.* **2008**, *43*, 7409–7417. [CrossRef]

14. Wielage, B.; Nickel, D.; Lampke, T.; Alisch, G.; Podlesak, H.; Darwich, S.; Hockauf, M. Corrosion characteristics of an ultrafine-grained Al-Mg-Si alloy (AA6082). *Mater. Sci. Forum* **2008**, *584–586*, 988–993.

15. Song, D.; Ma, A.B.; Jiang, J.H.; Lin, P.H.; Yang, D.H. Corrosion behavior of ultra-fine grained industrial pure Al fabricated by ECAP. *Trans. Nonferrous Metals Soc. China* **2009**, *19*, 1065–1070. [CrossRef]

16. Korchef, A.; Kahoul, A. Corrosion Behavior of Commercial Aluminum Alloy Processed by Equal Channel Angular Pressing. *Inter. J. Corros.* **2013**. [CrossRef]

17. Jilani, O.; Njah, N.; Ponthiaux, P. Transition from intergranular to pitting corrosion in fine grained aluminum processed by equal channel angular pressing. *Corros. Sci.* **2014**, *87*, 259–264. [CrossRef]

18. Naeini, M.F.; Shariat, M.H.; Eizadjou, M. On the chloride induced pitting of ultra fine grains 5052 aluminum alloy produced by accumulative roll bonding process. *J. Alloys Compd.* **2011**, *509*, 4696–4700. [CrossRef]

19. Wei, W.; Wei, K.X.; Du, Q.B. Corrosion and tensile behaviors of ultra-fine grained Al–Mn alloy produced by accumulative roll bonding. *Mater. Sci. Eng. A* **2007**, *454*, 536–541. [CrossRef]

20. Suter, T.; Peter, T.; Böhni, H. Microelectrochemical investigations of MnS inclusions. *Mater. Sci. Forum* **1995**, *192–194*, 25–39.

21. Yulinova, A.; Nickel, D.; Frint, P.; Lampke, T. Electrochemical properties of AL-6060 alloy after industrial scale ECAP. *Mater. Sci.* **2012**, *48*, 191–196. [CrossRef]

22. Frint, P.; Hockauf, M.; Halle, T.; Strehl, G.; Lampke, T.; Wagner, M.F.X. Microstructural features and mechanical properties after industrial scale ECAP of an Al-6060 alloy. *Mater. Sci. Forum* **2011**, *667–669*, 1153–1158.

23. Frint, P.; Hockauf, M.; Dietrich, D.; Halle, T.; Wagner, M.F.X.; Lampke, T. Influence of strain gradients on the grain refinement during industrial scale ECAP. *Mater. Sci. Eng. Technol.* **2011**, *42*, 680–685.

24. Iwahashi, Y.; Wang, J.T.; Horita, Z.; Nemoto, M.; Langdon, T.G. Principle of equal-channel angular pressing for the processing of ultra-fine grained materials. *Scripta Mater.* **1996**, *35*, 143–146. [CrossRef]

25. Dietrich, D.; Berek, H.; Schulze, A.; Scharf, I.; Lampke, T. EBSD and STEM on aluminium alloys subjected to severe plastic deformation. *Prac. Metall.* **2011**, *43*, 136–150. [CrossRef]

26. Böhni, H.; Suter, T.; Schreyer, A. Micro- and nanotechniques to study localized corrosion. *Electrochim. Acta* **1995**, *40*, 1361–1368. [CrossRef]

27. Birbilis, N.; Buchheit, R.G. Electrochemical characteristics of intermetallic phases in aluminum alloys—An experimental survey and discussion. *J. Electrochem. Soc.* **2005**, *152*, B140–B151. [CrossRef]

28. Couper, M.J.; Rinderer, B.; Yao, J.Y. Characterisation of AlFeSi intermetallics in 6000 series aluminium alloy extrusions. *Mater. Sci. Forum* **2006**, *519–521*, 303–308.

29. Milkereit, B.; Wanderka, N.; Schick, C.; Kessler, O. Continuous cooling precipitation diagrams of Al-Mg–Si alloys. *Mater. Sci. Eng.* **2012**, *A550*, 87–96. [CrossRef]

Metals **2015**, *5*, 172–191

30. Birol, Y. Precipitation during homogenization cooling in AlMgSi alloys. *Trans. Nonferrous Metals Soc. China* **2013**, *23*, 1875–1881. [CrossRef]

31. Bournane, M.; Berezina, A.; Davydenko, O.; Monastyrska, T.; Molebny, O.; Spuskanyuk, V.; Kotko, A. Effect of severe plastic deformation on structure and properties of Al-Mg-Si alloy of 6060 type. *Mater. Sci. Metall. Eng.* **2013**, *1*, 13–21.

32. Hockauf, K.; Wagner, M.F.X.; Halle, T.; Niendorf, T.; Hockauf, M.; Lampke, T. Influence of precipitates on low-cycle fatigue and crack growth behavior in an ultrafine-grained aluminum alloy. *Acta Mater.* **2014**, *80*, 250–263. [CrossRef]

33. Suter, T.; Alkire, R.C. Microelectrochemical studies of pit initiation at single inclusions in Al 2024-T3. *J. Electrochem. Soc.* **2001**, *148*, B36–B42.

34. Wloka, J.; Hack, T.; Virtanen, S. Local electrochemical properties of laser beam-welded high-strength Al–Zn–Mg–Cu alloys. *Mater. Corros.* **2008**, *59*, 5–13.

Chapter 5:
UFG Copper Alloys

metals

MDPI

Article

Ultrafine-Grained Precipitation Hardened Copper Alloys by Swaging or Accumulative Roll Bonding

Igor Altenberger [1], Hans-Achim Kuhn [1,*], Mozhgan Gholami [2], Mansour Mhaede [2] and Lothar Wagner [2]

1   Central Laboratory, Research & Development, Wieland-Werke AG, Graf-Arco-Str. 36, 89079 Ulm, Germany; igor.altenberger@wieland.de

2   Institute of Materials Science and Engineering, TU Clausthal, Agricola-Str. 6, 38678 Clausthal-Zellerfeld, Germany; mozhgan.gholami.kermanshahi@tu-clausthal.de (M.G.); mansour.mhaede@tu-clausthal.de (M.M.); lothar.wagner@tu-clausthal.de (L.W.)

*   Author to whom correspondence should be addressed; achim.kuhn@wieland.de; Tel.: +49-0-731-944-3705.

Academic Editor: Heinz Werner Höppel
Received: 30 March 2015; Accepted: 8 May 2015; Published: 13 May 2015

Abstract: There is an increasing demand in the industry for conductive high strength copper alloys. Traditionally, alloy systems capable of precipitation hardening have been the first choice for electromechanical connector materials. Recently, ultrafine-grained materials have gained enormous attention in the materials science community as well as in first industrial applications (see, for instance, proceedings of NANO SPD conferences). In this study the potential of precipitation hardened ultra-fine grained copper alloys is outlined and discussed. For this purpose, swaging or accumulative roll-bonding is applied to typical precipitation hardened high-strength copper alloys such as Corson alloys. A detailed description of the microstructure is given by means of EBSD, Electron Channeling Imaging (ECCI) methods and consequences for mechanical properties (tensile strength as well as fatigue) and electrical conductivity are discussed. Finally the role of precipitates for thermal stability is investigated and promising concepts (e.g. tailoring of stacking fault energy for grain size reduction) and alloy systems for the future are proposed and discussed. The relation between electrical conductivity and strength is reported.

Keywords: Cu-Ni-Si alloys; swaging; accumulative roll bonding; precipitation hardening

1. Introduction

Precipitation hardening can provide a combination of high strength and high (thermal or electrical) conductivity in copper alloys. By concentrating the alloying elements in fine precipitates, the Cu-matrix remains relatively pure with only few interstitial or substitutional atoms left in the Cu-matrix. Consequently conductivity is not detrimentally affected by solid solution impurities while maintaining high yield strength by finely dispersed precipitates which effectively impede dislocation movement.

The industrially most relevant precipitation hardened copper alloys combining high strength with high electrical conductivity are essentially Corson-alloys [1] which are often (but not exclusively) based on the ternary system Cu-Ni-Si. The Cu-Ni-Si-system has been thoroughly studied already in 1927 [1]. Microstructurally, the high yield strength (up to 800–900 MPa after precipitation hardening and cold working) is caused by finely dispersed semi-coherent Ni-Si-precipitates [2,3] with a diameter lower than 20 nm. As explained by the phase diagram (Figure 1), the temperature of solution annealing strongly depends on the amount of alloyed Ni and Si. Today, preferred and standardized alloys such as C7025 or C7035 contain ~3% silicides.

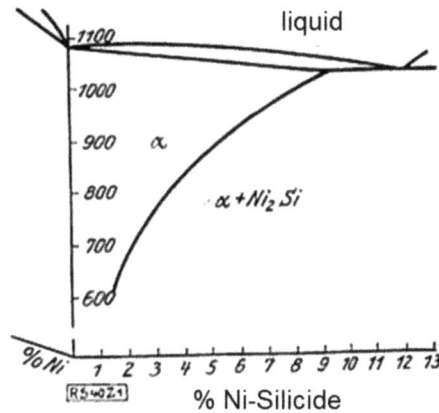

Figure 1. Pseudo-binary phase diagram of the Cu-Ni$_2$Si-system after Corson, 1927 [1].

The traditional approach for the development of high strength copper alloys is focused on chemical variation of precipitation hardened alloys [4]. In addition, microstructural control, e.g., generation of very fine grained or even ultra fine grained copper alloys, opens the door for tailored copper alloys combining optimized precipitate- as well as grain- or subgrain structure. Both approaches are presently used to generate high-performance components for industrial practice. Common methods for generating ultra fine grained metals by Severe Plastic Deformation (SPD), such as Equal Channel Angular Pressing (ECAP) or Accumulative Roll Bonding (ARB) [5–7], are well known and established, especially for pure copper. In contrast to SPD-related studies on pure copper, the archival literature sources dealing with severe plastic deformation of copper *alloys* are significantly more rare [8,9].

In the present study the authors seek to investigate and discuss the applicability of swaging (as a continuous method) as well as ARB for achieving very fine grained to ultra fine grained microstructures in classical Cu-Ni-Si alloys. A key feature of the research presented here is the stabilization of the microstructure by optimized aging treatments after swaging or ARB.

2. Experimental Section

The investigated copper alloy is the Corson-type alloy CuNi3Si1Mg (UNS designation C70250), which has experienced wide-spread use as connector-, leadframe- and high-strength wire material. Traces of Mg are alloyed to enhance the stress relaxation stability. Mg contributes to solid solution hardening as well as to precipitation hardening since Mg atoms may also form mixed (Ni,Mg)-silicides. The material investigated in our present study was hot extruded at 900 °C, then solution annealed at 800 °C/2 h (or alternatively at 950 °C/10 min). This condition was then rotary swaged and finally precipitation hardened. It should be noted that a complete dissolution of coarse Ni-silicides is not possible at these temperatures [10]. In the present study, the precipitation hardening was carried out at 450 °C.

Backscatter (ECCI, electron channeling contrast imaging) electron microscopy [11,12] was carried out using an AsB (Angle Selective Detector, Zeiss, Oberkochen, Germany) [13] in a Zeiss ULTRA scanning electron microscope (SEM, Zeiss, Oberkochen, Germany) equipped with a thermal field emission cathode. Typically, an aperture lens of 120 μm and acceleration voltages of 15–20 kV at a working distance of 2–6 mm were used.

For the Electron Backscatter Diffraction (EBSD) investigations, an EBSD-unit by Oxford was used. The EBSD patterns were recorded using a 4 × 4 binning, data acquisition and calculation of the patterns were performed by a Nordlys camera and AZTEC software by Oxford (UK), respectively. Prior to EBSD and ECCI-characterization in the SEM, the samples were carefully mechanically ground up to

2400 grid (SiC paper) and then polished up to 1 μm. Finally, samples were vibration polished for 3 h with dispersed magnesium oxide to aim for a sample surface with as little preparational cold work as possible. Two sets of experiments were carried out: swaging (which was carried out at TU Clausthal, Clausthal-Zellerfeld, Germany) of solution annealed bars from an initial diameter of 24 mm to a diameter of 7 mm (phi = −2.4) as well as swaging of a solution annealed wire with a diameter of 5.3 mm to a diameter of 2.7 mm (phi = −1.39). In both cases, precipitation annealing after swaging was done at 450 °C at different aging times. In the following elaborations we will use the terminology "peak-aged" for samples which were precipitation hardened at 450 °C for 1–6 h and "over-aged" for samples which were precipitation hardened for >16 h. Further details concerning the aging kinetics for the 2.7 mm wire are given in [14]. Finally, the swaged and subsequently precipitation hardened samples were mechanically characterized by tensile- and hardness tests. Moreover, the electrical conductivity of all the samples, before and after artificial aging, was measured using a SIGMATEST®-probe (eddy current method, Foerster, Reutlingen, Germany).

3. Results and Discussion

CuNi3Si1Mg was hot extruded at 800–900 °C, then solution treated at 800 °C/2 h (or alternatively 950 °C/10 min) and subsequently water-quenched. After this treatment the alloy exhibited a coarse grained microstructure with grain sizes of 100–150 μm and a few coarse silicides (typical diameter of a few hundred nm) which were not completely dissolved during homogenization.

The strain hardening curve for the solution treated and swaged condition can be seen in Figure 2. The highest hardness increase was observed after swaging up to a logarithmic strain of 2.5, then some saturation or even slight softening takes place. Therefore, further investigations focused on the condition swaged to a strain of 2.5, corresponding to a hardness increase of 100 HV as compared to the solution annealed state.

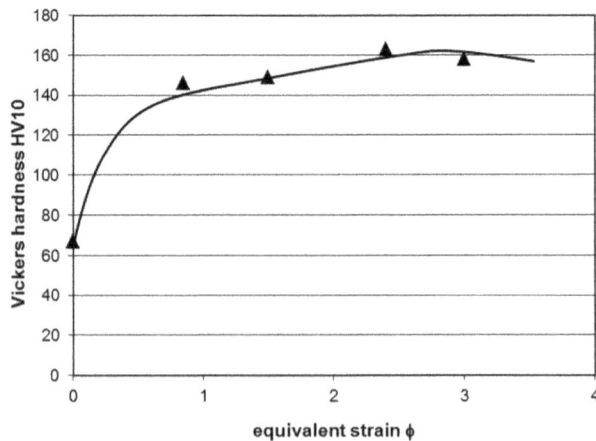

Figure 2. Hardness evolution of solution treated CuNi3Si1Mg by swaging.

After the extruded and solution treated bars were swaged from a diameter of 24 mm down to a diameter of 7 mm, an artificial aging treatment at 450 °C for aging times ranging from 30 min to 16 h was carried out. Figure 3 exhibits the resulting ultra fine grained microstructure at high resolution (magnification ~10000 times) before and after aging (distance from surface 100 μm). The swaging treatment significantly reduced the grain- or subgrain size to 200–800 nm. In addition, the formed ultra-fine grains show some pronounced elongation. With increasing distance from surface the grains become more equiaxed and their size increased up to ~2 μm in the center of the swaged bar. With increasing aging time, characteristic structural changes can be observed in the grain- and precipitate size- and arrangement (Figure 3). More and more precipitates (presumably largely Ni_2Si [10]) are formed as the aging continues, until, after 5–6 h at 450 °C a maximum hardness increase (peak-aging) occurs (see also Figures 4 and 6). By then, the grains have become more equiaxed in shape, however the grain coarsening is still not very pronounced. The precipitates are mainly concentrated on grain boundaries (Figure 3, 5 h). Only after extended aging of 16 h at 450 °C (Figure 3, 16 h) the microstructure becomes over-aged with clearly coarser grains as well as coarse precipitates, leading to a decline of hardness also. In this context, the characteristic sizes of the precipitates in the peak-aged and over-aged condition have to be pointed out.

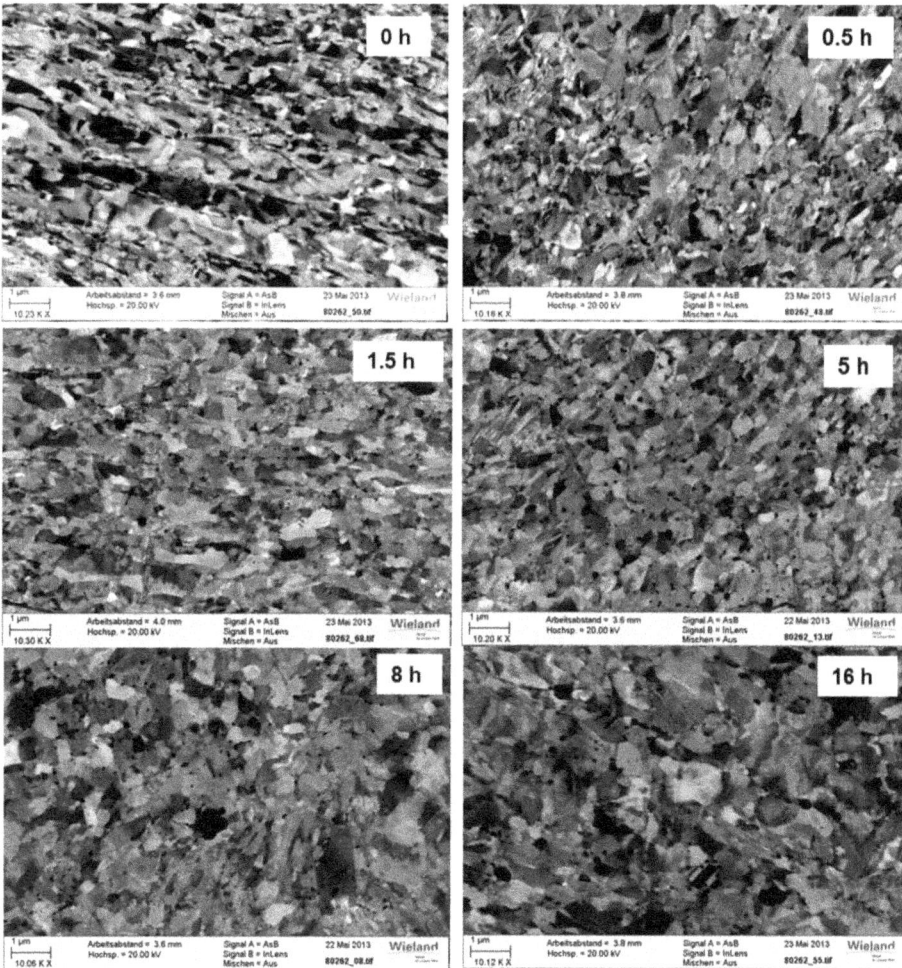

Figure 3. Microstructure of swaged CuNi3Si1Mg after different aging times ($T = 450$ °C) (ECCI-micrographs).

Figures 4 and 5 illustrate the size distribution of the precipitates in the peak-aged and over-aged condition, as derived from manual counting and measuring of precipitates in Electron Channeling Contrast Imaging (ECCI)–SEM micrographs. At the hardness peak (peak-aging) the average precipitate diameter is around 14 nm, whereas in the over-aged condition, after 16 h, the average precipitate diameter increased to around 47 nm. These aging kinetics are 3 to 5 times faster than in the non-swaged condition where a hardness maximum is found after 16 hours accompanied by typical mean precipitate diameters of 5 nm (see also [2] for comparison). Interestingly, for both conditions, peak-aged as well as over-aged, a bimodal precipitate size distribution was detected (Figures 4 and 5). By means of Energy Dispersive Spectroscopy (EDS) microanalysis we can not distinguish between nanoscale orthorhombic Ni_2Si and the possible hexagonal minority phase $Ni_{31}Si_{12}$ (which is thermodynamically expected for low Ni-content of ~2%). From measurements using EBSD, it appeared that there is also a fraction of hexagonal Ni-silicides. If this observation is any evidence of $Ni_{31}Si_{12}$-phase in addition to Ni_2Si, it is likely that the bimodal size distribution is also driven by this second type of precipitation. Alternatively, it can also be speculated, that the bimodal size distribution is possibly caused by different aging kinetics of precipitates at or near grain/subgrain boundaries and within the grains where diffusion is drastically different.

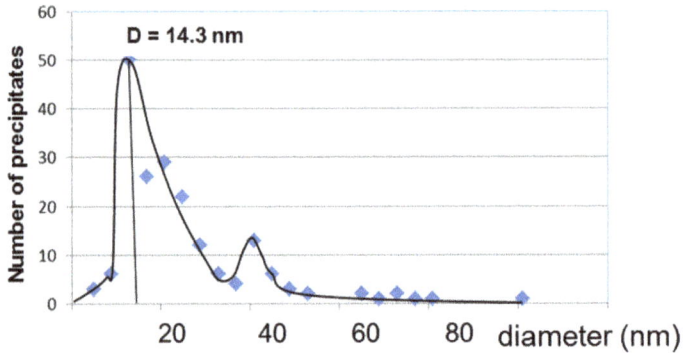

Figure 4. Precipitate size distribution in the peak-aged condition.

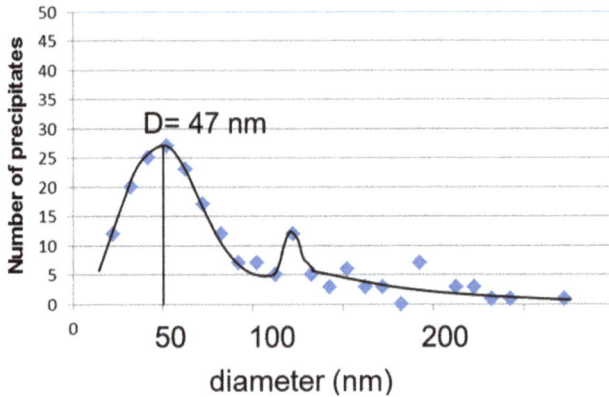

Figure 5. Precipitate size distribution in the over-aged condition.

Figure 6 exhibits the aging curves (hardness *vs.* aging time) of swaged CuNi3Si1Mg for isothermal aging in the temperature range 300–500 °C. Hardness values of up to 245 HV can be reached in the peak-aged conditions at aging temperatures of 400, 450 or 500 °C. With increasing temperature the hardness peak is shifted to smaller aging times. At an aging temperature of 350 °C, only a maximum hardness of ~230 HV appears to be possible.

Figure 7 shows the change of electrical conductivity of the swaged and non-swaged condition during aging at 450 °C. The diffusion of the alloying elements Ni and Si from the solid solution into the precipitates decreases the scattering of electrons by the strain fields of solute atoms. As a result, the electrical conductivity increases.

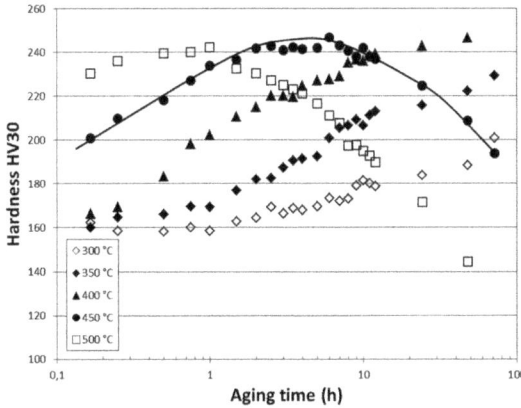

Figure 6. Aging curves of swaged CuNi3Si1Mg for different aging temperatures.

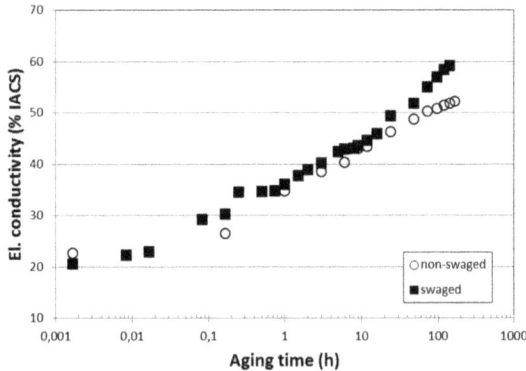

Figure 7. Conductivity *vs.* aging time for swaged (UFG) and coarse-grained CuNi3Si1Mg (*T* = 450 °C).

It is noteworthy, that already after 10 min the conductivity of the swaged condition is slightly higher than the conductivity of the non-deformed condition at this temperature. Obviously, the diffusion of solute elements is significantly accelerated by fast diffusion paths such as high- and low angle boundaries which are prevalent in the swaged condition. Throughout the further aging process the electrical conductivity of the swaged condition stays superior to the conductivity of the non-deformed condition. This difference amounts up to 8% IACS (International Annealed Copper Standard, 58 MS/m) in severely over-aged specimens, possibly being caused also by recrystallization which drastically reduces the grain boundary area in the swaged and severely over-aged condition. For comparison, standardized commercial CuNi3Si1Mg strips typically have electrical conductivities of 35%–45% IACS.

At an aging temperature of 450 °C, the microstructure of the swaged condition was not stable for long aging times. Nevertheless, at a lower aging temperature of 300 °C an aging effect can be induced without pronounced over-aging. At this temperature, a drop of hardness is not observed within several hundred hours thermal exposure (Figure 8). For the application of electromechanical connectors, this is a significant finding, since electromechanical connectors (in the presented conductivity range) are usually not exposed to temperatures higher than 150–200 °C during service. At this moment, implications for stress relaxation behavior of swaged or severely deformed precipitation hardened Corson-alloys remain speculative, however it can be assumed, that finely precipitated Ni-silicides at grain boundaries as well as within grains may also serve to effectively diminish stress relaxation or creep.

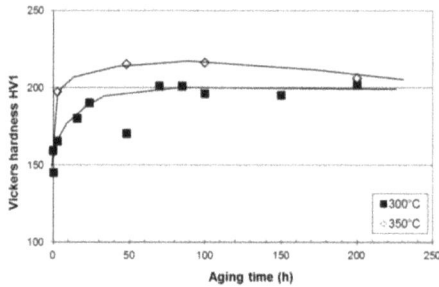

Figure 8. Hardness evolution of swaged CuNi3Si1Mg during thermal exposure at 300 °C (specimens swaged from 5.3 to 2.7 mm diameter).

As aforementioned, no homogeneous ultrafine grained (UFG) structure was achieved by swaging CuNi3Si1Mg bars from diameters of 24 mm to 7 mm. However, for wire which was swaged from 5.3 to 2.7 mm, a fully ultra fine grained structure was observed in the whole cross section after swaging and subsequent aging. Figures 9 and 10 show ECCI- as well as EBSD results for the obtained microstructure after aging at 450 °C for one hour. This microstructure (pictured in the center of the specimen) is characterized by strong orientation contrast (Figure 9), rather low dislocation density within the grains and a high percentage of high-angle boundaries (as seen by mapping grain orientations (Figure 10a) and grain boundary misorientations (Figure 10b) using EBSD). The typical resulting grain size in this UFG structure is about 350 nm.

Figure 9. ECCI-picture of ultra-fine grained CuNi3Si1MgSi (swaged and aged at 450 °C/1 h, swaging from diameter of 5.3 to 2.7 mm).

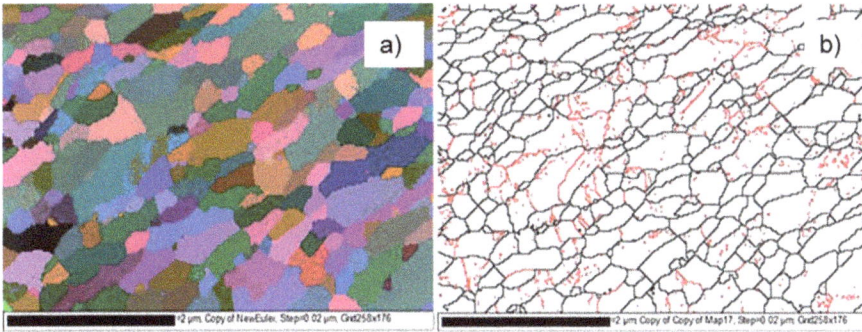

Figure 10. Fully ultra-fine grained CuNi3Si1Mg Electron Backscatter Diffraction Pattern (EBSD) orientation mapping showing Euler-orientations (**a**) and grain boundaries (**b**). High-angle grain boundaries are depicted in black, Low-angle grain boundaries are depicted in red).

Ultra fine grained microstructures can be obtained also in CuNi3Si1Mg by accumulative roll bonding (ARB) [15]. Figure 11 shows a high resolution ECCI micrograph of ARB-processed CuNi3Si1Mg. (for details see [15]). The typical microstructure of ARB-processed CuNi3Si1Mg is characterized by elongated grains perpendicular to rolling direction, having grain diameters lower than 200 nm. An analysis of high angle grain boundaries by EBSD reveals a grain width of 100 nm. This corresponds to other findings [16], where for the same logarithmic strain of ~5, ARB-processed pure copper showed similar grain widths. Also here, aging experiments after severe plastic deformation were carried out. In analogy to the swaged condition, precipitates were formed preferentially at grain boundaries, thus reducing the grain boundary mobility during further thermal exposure.

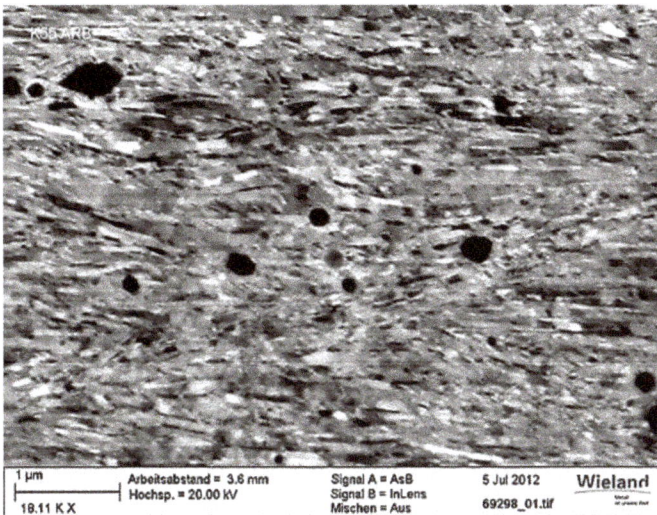

Figure 11. Microstructure of solution treated ARB-processed CuNi3Si1Mg. The larger precipitates were not dissolved by the solution treatment and are remains from the prematerial as produced by continuous casting.

Figure 12 exhibits a thermal softening curve (hardness vs. temperature) of differently processed CuNi3Si1Mg after 1 hour annealing at 300 to 600 °C. If we define the onset of softening as a loss of 10% in initial hardness, the conventionally processed spring hard strip softens at 500 °C whereas the swaged and ARB conditions start softening at 475 °C. This thermal stability is significantly higher compared to pure copper, where softening in the ARB condition already starts at 200 °C [17]. By using high resolution ECCI, grain boundary pinning by Ni$_2$Si-precipitates is confirmed as the underlying mechanism for the high thermal stability (Figure 13).

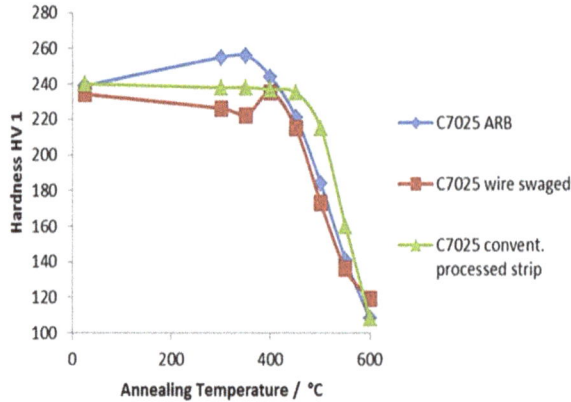

Figure 12. Thermal stability of swaged and ARB-processed CuNi3Si1Mg as compared to conventional strip.

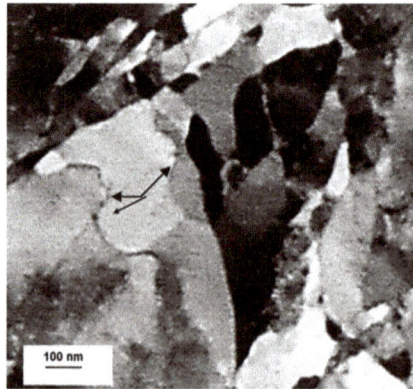

Figure 13. ECCI-micrograph of swaged and under-aged CuNi3Si1Mg showing a low-angle grain boundary which is pinned by Ni-silicides (arrows).

The resulting mechanical (quasistatic as well as cyclic) properties after severe plastic deformation and optimized aging treatments are significantly superior to conventionally processed strip. As an example, thin CuNi3Si1Mg wires (diameter 0.1 mm) which were processed from swaged and optimized aged bars show tensile strengths of up to 1050 MPa and yield strengths higher than 1000 MPa. However, this strength increase is achieved at the expensive of electrical conductivity which is reduced to less than 25% IACS.

In addition to excellent quasistatic strength, optimized swaging plus consecutive aging leads to a marked increase of the 10^7-fatigue endurance strength from 250 to 300 MPa (Figure 14). By combining these processes with a final mechanical surface treatment (such as shot peening, laser peening or deep rolling) even higher fatigue strengths of about 400 MPa can be achieved [18]. Higher fatigue strengths for copper alloys are only reported for high-alloyed spray-formed copper alloys and Cu-Be alloys [19].

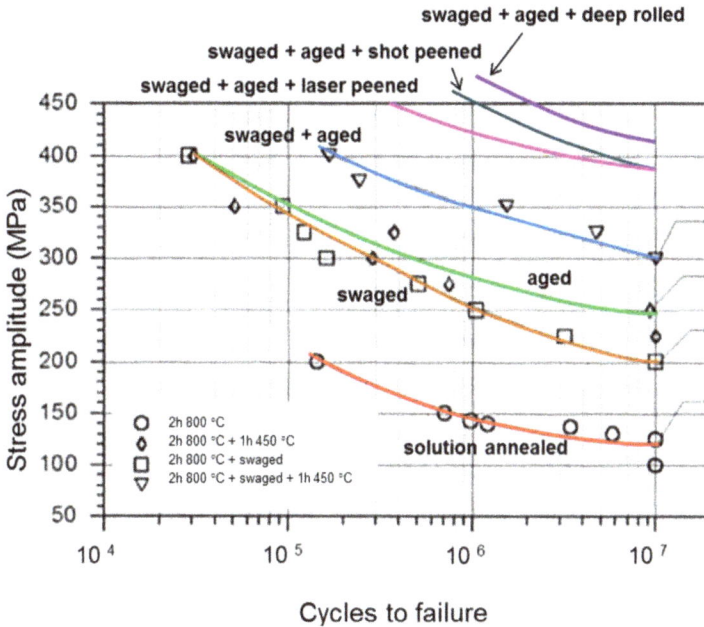

Figure 14. Wöhler-curves (rotation bending, R = −1) of swaged and non-swaged optimized precipitation hardened CuNi3Si1Mg for different mechanical surface treatments (part of this data from [18]).

As an outlook, in context with swaging the following further approaches should be addressed:

Firstly, CuNi3Si1Mg is low alloyed, therefore the achievable strength is limited by the volume fraction of the precipitates. To enable highest strengths of CuNi3Si1Mg, significant cold work is necessary (e.g., by drawing) which in turn lowers the conductivity. Higher alloyed Cu-Ni-Si alloys or more complex alloy systems such Cu-Ni-Si-Cr or Cu-Ni-Co-Si allow higher volume fractions of precipitates or multiphase hardening, respectively. First results on swaged and aged CuNi7Si2Cr are promising and indicate possible conductivities of 34% IACS at tensile strengths of ~1100 MPa after swaging and optimized aging, which is in the strength regime of Cu-Be-alloys.

Another approach to generate UFG copper alloys is to introduce an extremely high twinning density by severe deforming of single-phased copper alloys which exhibit very low stacking fault energies (such as Cu-Al, Cu-Al-Si, Cu-Zn or Cu-Zn-Si). By finely spaced twinning a very fine grain size can be achieved after swaging leading to a hardness of more than 420 HV (which corresponds to tensile strengths higher than 1300 MPa). Figure 15 shows the refinement of the microstructure by twinning in CuAl5.8Si2 (estimated stacking fault energy around 6 mJ/m^2 or lower [20,21]) with increasing logarithmic deformation degree during swaging. However, CuAl5.8Si2 is non-age hardenable, therefore there are no precipitates stabilizing the grain boundaries during thermal exposure. As a consequence, a hardness loss of more than 10% is observed at 300 °C after just one hour.

Metals **2015**, *5*, 763–776

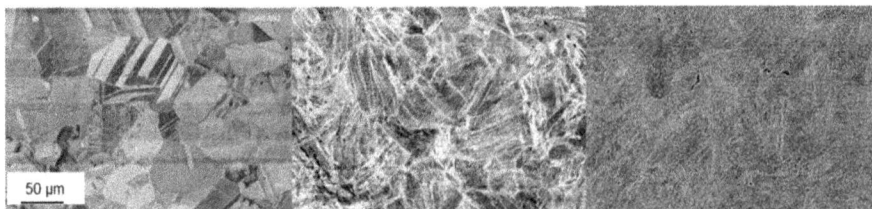

Figure 15. Evolution of an ultrafine-grained structure in single phased low-stacking fault alloy CuAl5.8Si2 by swaging (**Left**: non-swaged; **Middle**: swaged to a logarithmic strain of −0.8; **Right**: swaged to a logarithmic strain of −1.4).

4. Conclusions

The combination of swaging and subsequent optimized precipitation hardening is a simple method to produce Cu-Ni-Si materials with very fine grain size in the range 0.2–2 μm. For strip material accumulative roll bonding (ARB) is a suitable method to generate ultra fine grained microstructures in Cu-Ni-Si alloys. A crucial role is ascribed to the artificial aging treatment after the severe plastic deformation. The thermal stability of ultra fine grained CuNi3Si1Mg is significantly enhanced as compared to pure copper, owing to nanoscopically small precipitates which effectively pin the grain boundaries during aging or annealing. At 300 °C no over-aging was detected within 200 h. For short time exposure (1 h) the grain structure is fairly stable up to 400 °C. In addition to enhanced ultimate tensile- and yield strength (with possible strength >1000 MPa), also the fatigue behavior in the High Cycle Fatigue (HCF)-regime was significantly improved by the UFG-structure in the swaged plus peak-aged condition close to the surface.

Acknowledgments: Experimental help during wire drawing of Cu-Ni-Si wire by D. Vucic-Seele is kindly acknowledged.

Author Contributions: I.A. is the primary author of the paper and performed analysis of the experimental data. M.G. and M.M. contributed to the experimental research work. H.A.K. and L.W. discussed the results and analysis with the other authors.

Conflicts of Interest: The authors declare no conflict of interest.

References

1. Corson, M.G. Copper hardened by a new method. *Z. Metallkunde* **1927**, *19*, 370–371.
2. Lockyer, S.A.; Noble, F.W. Precipitate structure in a Cu-Ni-Si alloy. *J. Mater. Sci.* **1994**, *29*, 218–226. [CrossRef]
3. Wang, C.; Zhu, J.; Lu, Y.; Guo, Y.; Liu, X. Thermodynamic description of the Cu-Ni-Si system. *J. Phase Equilib. Diffus.* **2014**, *35*, 93–104. [CrossRef]
4. Kuhn, H.-A.; Altenberger, I.; Käufler, A.; Hölzl, H.; Fünfer, M. Properties of high performance alloys for electromechanical connectors. In *Copper Alloys—Early Applications and Current Performance—Enhancing Processes*; Collini, L., Ed.; InTech: Rijeka, Croatia, 2012; p. 52.
5. Valiev, R.Z.; Islamgaliev, R.K.; Alexandrov, I.V. Bulk nanostructured materials from severe plastic deformation. *Prog. Mater. Sci.* **2000**, *45*, 103–189. [CrossRef]
6. Höppel, H.W.; May, J.; Göken, M. Enhanced strength and ductility in ultrafine-grained aluminium produced by accumulative roll bonding. *Adv. Eng. Mater.* **2004**, *6*, 781–784. [CrossRef]
7. Mughrabi, H.; Höppel, H.W.; Kautz, M. Fatigue and microstructure of ultrafine-grained metals produced by severe plastic deformation. *Scr. Mater.* **2004**, *51*, 807–812. [CrossRef]
8. Neishi, K.; Horita, Z.; Langdon, T.G. Achieving superplasticity in a Cu–40%Zn alloy through severe plastic deformation. *Scr. Mater.* **2001**, *45*, 965–970. [CrossRef]
9. Wang, J.; Zhang, P.; Duan, Q.; Yang, G.; Wu, S.; Zhang, Z. Tensile deformation behaviors of Cu-Ni alloy processed by equal channel angular pressing. *Adv. Eng. Mater.* **2010**, *12*, 304–311. [CrossRef]

10. Kinder, J.; Huter, D. TEM-Untersuchungen an höherfesten und elektrisch hochleitfähigen CuNi2Si-Legierungen. *Metall* **2009**, *63*, 298–303.

11. Altenberger, I.; Kuhn, H.-A.; Gholami, M.; Mhaede, M.; Wagner, L. Characterization of ultrafine grained Cu-Ni-Si alloys by electron backscatter diffraction. *IOP Conf. Ser. Mater. Sci. Eng.* **2014**, *63*, 012135. [CrossRef]

12. Altenberger, I.; Kuhn, H.-A.; Hölzl, H. Mikrostrukturelle Charakterisierung von hochfesten Cu-Ni-Si-Legierungen mittels Electron-Channeling-Rückstreukontrast im Rasterelektronenmikroskop. *Sonderband d. Prakt. Metallographie* **2012**, *44*, 79–84. (In German)

13. Jaksch, H. Strain related contrast mechanisms in crystalline materials imaged with AsB detection. In *EMC 2008-14th European Microscopy Congress 2008*; Luysberg, M., Tillmann, K., Weirich, T., Eds.; Springer: Berlin/Heidelberg, Germany, 2008; pp. 553–554.

14. Altenberger, I.; Kuhn, H.-A.; Mhaede, M.; Gholami, M.; Wagner, L. Wie viel NANO steckt in Kupfer–ein klassischer Werkstoff im 21. Jahrhundert. *Metall* **2012**, *66*, 500–504. (In German)

15. Kuhn, H.-A.; Altenberger, I.; Riedle, J.; Hölzl, H. Microstructure and mechanical properties of ultra fine grained high performance copper alloys. In Proceedings of Copper 2013; Leibbrandt, J., Ignat, M., Sanchez, M., Eds.; The Chilean Institute of Mining Engineers: Santiago, Chile, 2013; pp. 129–138.

16. Murata, Y.; Nakaya, I.; Morinaga, M. Assessment of strain energy by measuring dislocation density in copper and aluminium prepared by ECAP and ARB. *Mat. Trans.* **2008**, *49*, 20–23. [CrossRef]

17. Li, Y.J.; Zeng, X.H.; Blum, W. Transition from strengthening to softening by grain boundaries in ultrafine-grained Cu. *Acta Mater.* **2004**, *52*, 5009–5018. [CrossRef]

18. Gholami, M.; Altenberger, I.; Mhaede, M.; Sano, Y.; Wagner, L. Surface treatments to improve fatigue performance of age-hardenable CuNi3Si1Mg. In Proceedings of the 12th International Conference on shot Peening, Goslar, Germany, 15–18 September 2014; Wagner, L., Ed.; pp. 208–213.

19. Altenberger, I.; Kuhn, H.-A.; Müller, H.R.; Mhaede, M.; Gholami, M.; Wagner, L. Material properties of high-strength Beryllium-free copper alloys. *Int. J. Mater. Prod. Technol.* **2015**, *50*, 124–146. [CrossRef]

20. Rohatgi, A.; Vecchio, K.S.; Gray, G.T., III. The influence of stacking fault energy on the mechanical behavior of Cu and Cu-Al alloys: Deformation twinning, work hardening, and dynamic recovery. *Metall. Mater. Trans A* **2001**, *32*, 135–145. [CrossRef]

21. An, X.H.; Lin, Q.Y.; Wu, S.D.; Zhang, Z.F.; Figueiredo, R.B.; Gao, N.; Langdon, T.G. The influence of stacking fault energy on the mechanical properties of nanostructured Cu and Cu–Al alloys processed by high-pressure torsion. *Scr. Mater.* **2011**, *64*, 954–957. [CrossRef]

metals

MDPI

Article

Grain Refinement and High-Performance of Equal-Channel Angular Pressed Cu-Mg Alloy for Electrical Contact Wire

Aibin Ma *, Chengcheng Zhu, Jianqing Chen, Jinghua Jiang, Dan Song, Shizhan Ni and Qing He

College of Mechanics and Materials, Hohai University, Nanjing 210098, China; zccheng1986@163.com (C.Z.); chenjq@hhu.edu.cn (J.C.); jinghua-jiang@hhu.edu.cn (J.J.); songdancharls@hhu.edu.cn (D.S.); nishizhan1988@163.com (S.N.); hoking2011@sina.com (Q.H.)

* Author to whom correspondence should be addressed; aibin-ma@hhu.edu.cn; Tel.: +86-25-8378-7239; Fax: +86-25-8378-6046.

External Editor: Heinz Werner Höppel

Received: 29 October 2014; in revised form: 25 November 2014; Accepted: 3 December 2014; Published: 9 December 2014

Abstract: Multi-pass equal-channel angular pressing (EACP) was applied to produce ultrafine-grained (UFG) Cu-0.2wt%Mg alloy contact wire with high mechanical/electric performance, aim to overcome the catenary barrier of high-speed trains by maximizing the tension and improving the power delivery. Microstructure evolution and overall properties of the Cu-Mg alloy after different severe-plastic-deformation (SPD) routes were investigated by microscopic observation, tensile and electric tests. The results show that the Cu-Mg alloy after multi-pass ECAP at 473 K obtains ultrafine grains, higher strength and desired conductivity. More passes of ECAP leads to finer grains and higher strength, but increasing ECAP temperature significantly lower the strength increment of the UFG alloy. Grain refinement via continuous SPD processing can endow the Cu-Mg alloy superior strength and good conductivity characteristics, which are advantageous to high-speed electrification railway systems.

Keywords: severe-plastic-deformation (SPD); contact wire; Cu-Mg alloy; ultrafine grained (UFG); conductivity; strength

1. Introduction

With the rapid development of high-speed electric railway, mainly, advanced metallic materials used for contact wire are desired. At present, the task of increasing the speed of the trains ($\geq$300 km/h) makes high-performance copper contact wire become a research focus. In this situation, copper alloy are usually required to possess high tensile strength (above 550 MPa) and good conductivity (about 60% IACS), with the aim to overcome the catenary barrier of high-speed trains by maximizing the tension and improving the power delivery. As well known, high strength and good conductivity of metals are often mutually exclusive. It is important to note that some abnormal methods have been reported to achieve a good combination of strength and conductivity [1,2], but those methods are relatively complicated in commercial applications. Most of the reports were focused on adding small amounts of alloy elements (such as Cr, Ag, Zr, Nb, Co, *etc.*) to pure copper [3–6] or applying hardening processes (such as drawing or rolling) [6]. Those conventional strengthening methods induce various kinds of defects (dislocations, reinforcing phases, point defects, grain boundaries), raising electrical resistivity of Cu alloy because of the scattering of conducting electrons [7,8]. For example, a Cu-0.7%Cr-0.3%Fe alloy, after cold working (40% CW), plus aged at 450 °C for 90 min, had the ultimate tensile strength of 460 MPa and an electrical conductivity of about 67% IACS (International Annealed Copper Standard, 17.24×10^{-9} Ωm is defined as 100% IACS) [9].

In view of the deteriorative effect on the conductivity of copper, the alloying content and hardening degree should be restricted during the fabrication of high-strength and good-conductivity copper alloy. Grain refinement of single solid-solution copper alloy (e.g., Cu alloy containing a small amount of Mg) could be an effective way to overcome the contradiction. At present, China Railway Construction Electrification Bureau Group (Kang Yuan New Materials Co., Ltd.) has developed fine-grained Cu-0.4wt%Mg contact wire by Conform-process plus cold drawing. Figure 1 presents the optical microstructure of the Cu-0.4wt%Mg alloy after Conform-process and subsequent cold drawing. The product achieved a good combination of ultimate tensile strength (UTS: 522 MPa) and conductivity (68.6% IACS), and was successfully applied in the high-speed railways of Zhengzhou-Xi'an and Korea. Recently, a novel severe-plastic deformation (SPD) procedure, namely equal-channel angular pressing (ECAP), has been applied for obtaining ultrafine-grained (UFG) copper alloy with high strength, large ductility, and good electrical conductivity [10–12]. In the present work, the SPD process was combined with the manufacturing processes of Cu-Mg alloy contact wire with a lower Mg content (about 0.2 wt% Mg), with the aim to achieve further improvement in conductivity and strength. Under the same deformation conditions, the average grain size of the Cu-0.2wt%Mg alloy (in Figure 2a) is obviously finer than that of the Cu-0.4wt%Mg alloy (in Figure 1a). Until now, there have been few reports on mechanical and conductivity properties of UFG Cu-Mg alloys fabricated by SPD methods. Herein, this paper studies the microstructure change, tensile strength, and electrical conductivity of the on-line conformed Cu-Mg binary alloy subjected to experimental multi-pass ECAP, and investigates the origin of the good characteristics.

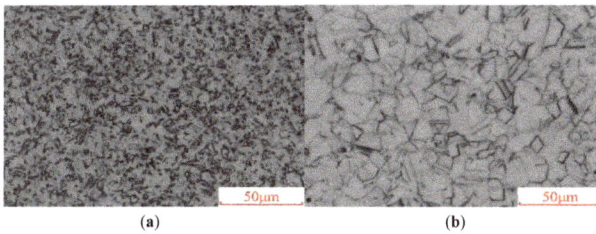

Figure 1. Optical micrographs of Cu-0.4wt%Mg alloy by Conform-process (**a**) and subsequent cold drawing (**b**) for contact wire products.

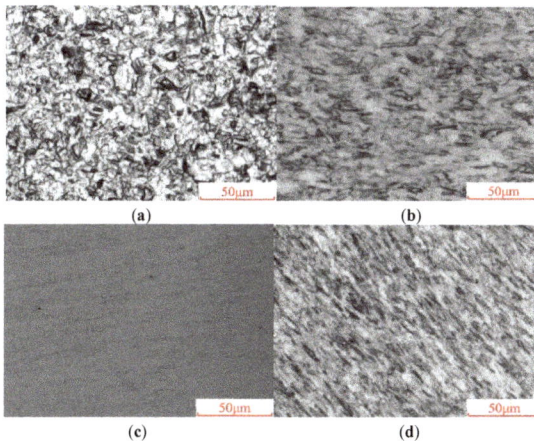

Figure 2. OM (Optical Micrographs) of Cu-0.2wt%Mg alloy (**a**) in Conform state and after ECAP at 473 K for (**b**) 1 pass, (**c**) 4 passes and (**d**) 16 passes.

2. Experimental Section

The material used was Cu-0.2wt%Mg (oxygen $\leq$10 ppm, 1 ppm = 10^{-6}) alloy, which was prepared by the upward-casting and then extruded by Conform-process in China railway construction electrification bureau group (Kang Yuan New Materials Co., Ltd., Jiangyin, China). Mg atoms of the binary alloy mainly exist in the FCC (Face-Center-Cubic)-structured copper crystal. The dimension of the ECAP billet was 19.5 mm × 19.5 mm × 40 mm. The ECAP process was carried out using a rotary die with an intersection angle of 90°, which details are described in previous references [13,14]. The billets were extruded from 1 pass to 16 passes at 473 K and 673 K, respectively. An Olympus BX51M optical microscope (OM, Olympus Corporation, Tokyo, Japan) was used to observe the microstructure at the flow plane of the ECAPed billets. The composition of the etchant was glacial acetic acid 25 mL, phosphoric acid 55 mL, and nitric acid 20 mL. The etching time was 5 s. A JEM-2000EX transmission electron microscope (TEM, JEOL Ltd., Tokyo, Japan) was applied to observe the microstructure and the grain size of the Cu-Mg alloy after ECAP.

Microhardness of the ECAPed sample was measured by HXD-1000TC device (Taiming Optical Instrument Co., Ltd., Shanghai, China) under a load of 100 g for 15 s. Tensile specimen with a dimension of 3 mm × 3 mm in cross-section, and 15 mm in gage length, was cut from the billet along the longitudinal direction. Tensile tests were performed by a RGM-4050 testing machine at an initial strain rate of 1000 μm/s at room temperature, and three samples cut from one billet were tested for each state. After tensile testing, the fracture surfaces were observed by a HITACHI S-3400N scanning electron microscope (SEM, HITACHI Ltd., Tokyo, Japan).

A QJ36S digital apparatus (Shuangte Electrical Instrument Co., Ltd., Shanghai, China) was implemented for the direct current (DC) electrical resistance measurement via a four-point probe method. The test samples with a dimension of 3 mm × 3 mm × 3 mm were cut from the billets along the longitudinal direction, and their surfaces were polished before the electrical resistance measurements.

3. Results and Discussion

3.1. Microstructure

Figure 2 presents the optical micrographs of the Conformed (as-achieved) alloy and the ones after ECAP at 473 K for different passes. As shown in Figure 2a, the average grain size of the Conformed sample was about 5–8 μm. There are only equiaxed α-Cu grains in microstructure of the alloy. After continuous ECAP processing, the grains were gradually refined and elongated with increasing the number of ECAP passes. The grain size was hard to measure when the sample was subjected to 16 passes of ECAP. Only fine strain-induced plastic flows can be observed by optical microscopy. The detailed microstructure of the 16-pass ECAPed copper alloy should be observed by TEM.

Figure 3 presents optical micrographs of the Cu-0.2wt%Mg alloy after ECAP at 673 K for one pass, four passes and 16 passes, respectively. Similar to the samples after ECAP at 473 K, the α-Cu grains were fined and elongated, however, the efficiency of grain refinement during ECAP at 673 K was not as good as that at 473 K. This phenomenon can be contributed to the faster grain recovery and recrystallization at 673 K [15]. Compared with the current product of Cu-0.4wt%Mg contact wires processed by Conform plus cold-drawing (in Figure 1), the grains of the Cu-0.2wt%Mg after Conform plus multi-pass ECAP are obviously finer.

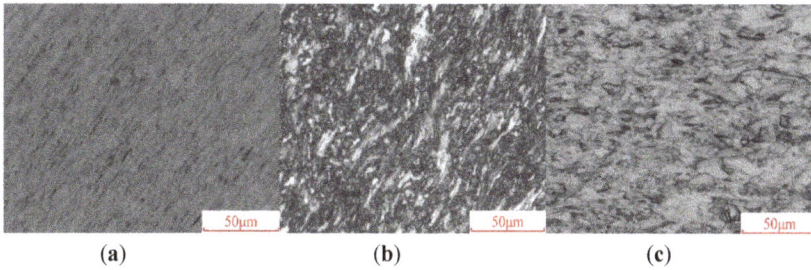

(a) **(b)** **(c)**

Figure 3. OM micrographs of the Cu-0.2wt%Mg alloy after ECAP at 673 K for (**a**) 1 pass, (**b**) 4 passes and (**c**) 16 passes.

Figure 4 presents TEM micrographs of Cu-0.2wt%Mg alloy after 16 passes of ECAP at 473 K. It can be seen that Mg element is completely dissolved into the Cu matrix and the dislocation density in the ECAPed alloy is pretty low (Figure 4a,b). Meanwhile, the grain size has already been reduced to about 200 nm. The corresponding SAD patterns (upper-right insertion of Figure 4a) are almost continuous diffraction rings, indicating the existence of a majority of high angle grain boundaries. Figure 4c presents dislocation cell structure in the ECAPed alloy. These cells may form individual subgrains upon further plastic straining. As well known, dislocation tangling is frequently observed in the interior of grains, where the grain is heavily strained [16]. There are also some nano-twins in particular grains, as can be seen in Figure 4d–f. Nano-twins were created by the shear stress and severe strain during the ECAP process.

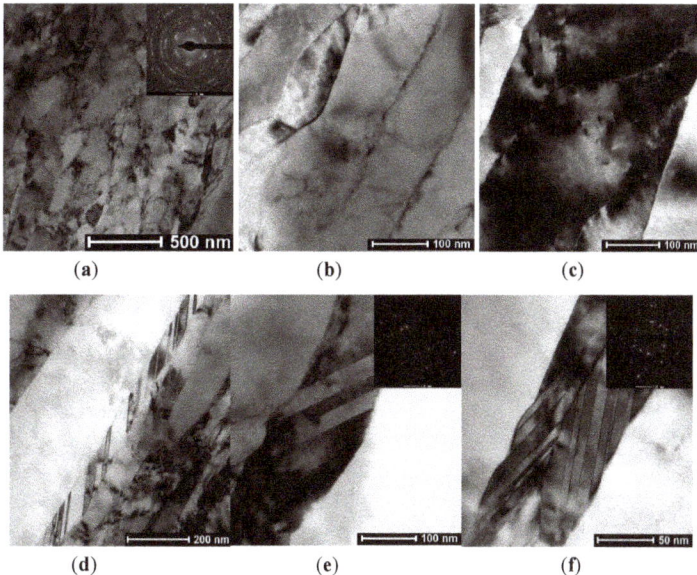

(a) **(b)** **(c)**

(d) **(e)** **(f)**

Figure 4. TEM microstructure of the ECAPed Cu-0.2wt%Mg alloy after 16 passes at 473 K. (**a**) Elongated ultra-fine grains observed at low magnification, (**b**) Elongated ultra-fine grains observed at high magnification, (**c**) characteristics of grain boundaries, (**d**) twins and intragranular dislocations, (**e**) characteristics of twins observed at high magnification and (**f**) characteristics of twins observed at the higher magnification.

3.2. Microhardness

Figure 5 presents the Vickers microhardness of Cu-0.2wt%Mg alloy subjected to different passes of the ECAP processing. Firstly, it can be seen that all the samples ECAPed at 473 K have higher hardness values than those of the samples at 673 K with the same ECAP passes. This is due to more obvious strain hardening and grain refinement at the lower ECAP temperature. Secondly, the results show an obvious increase in hardness from the first pass to four passes. The rapid increase of hardness at initial passes seems to be attributed to strain hardening rather than grain refinement in the initial stage [17]. Thirdly, the hardness value of the sample ECAPed at 673 K has a slight increase with more passes. The reason might be that the materials reached the steady-state density of dislocation and dynamic recovery occurred in the grains of Cu-Mg alloy [18]. While the hardness value of the sample ECAPed at 473 K obviously increase after 16 passes. This could be due to the efficient grain refinement and the slight dynamic recovery happened at 473 K.

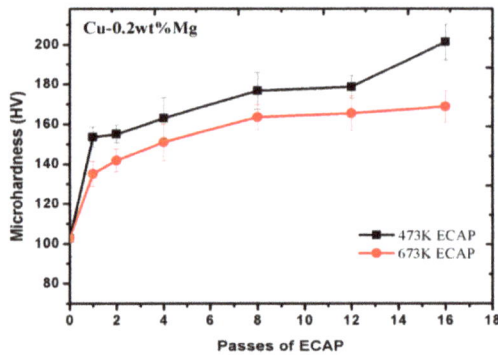

Figure 5. Microhardness of Cu-0.2wt%Mg alloy variation with the passes of ECAP at 473 K and 673 K.

3.3. Tensile Properties

Figure 6 shows engineering stress-strain curves of UFG Cu-0.2wt%Mg alloy processed by ECAP at different temperature. The as-achieved sample in Conform state shows high ductility and low strength, while the ECAPed samples exhibit much higher strength with adequate ductility. This might be caused by two reasons. Firstly, the grains in Conform state are relative large and equiaxed. A smaller grain size leads to a higher mechanical strength, which is widely known as the Hall-Petch effect. Secondly, the dislocation was annihilation in the dynamic recovery caused by the high temperature during Conform process. After multi-pass ECAP, the strength of the alloy was improved significantly, but the elongation was obviously decreased. The increase of strength attributes to strain hardening and grain refinement at a few passes. Ultrafine grains with high-angle grain boundaries impeded the motion of dislocations, which is the main reason for strength improvement of the ECAPed Cu-Mg alloy. Simultaneously, the twin lamellas in grains may act as barriers to largely reduce the dislocation mean free path, and, thus, further harden the alloy. Moreover, the strain hardening and the elongated grains caused by multi-pass ECAP reduced the ductility of the Cu-0.2wt%Mg alloy.

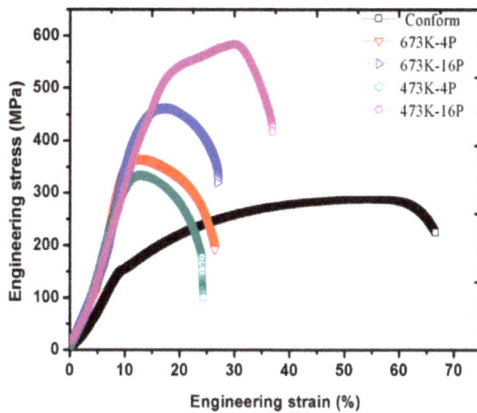

Figure 6. Stress-strain tensile curves of the samples subjected to various passes ECAP at 437 K and 673 K.

Table 1 lists microhardness and tensile properties of the samples subjected to ECAP at 473 K and 673 K. Firstly, the 16-pass samples exhibit higher strength and better ductility than the four-pass samples ECAPed at the two temperatures. This may be induced by the further refined grains and a more uniform distribution of grains. Secondly, the sample after 16 passes of EACP at 473 K obtained a higher strength and better ductility than that at 673 K. The 16-pass sample at 473 K results in a high tensile strength (583.4 MPa), good total elongation (37.9%), and high microhardness (201.23 HV). The improvement in strength may be caused by the obstruction of the large-angle grain boundaries and twin-grain boundaries to the dislocation movement. The better ductility may be induced by the deformation mechanism change from dislocation slip to grain-boundary sliding (GBS) [15].

Table 1. Microhardness and tensile properties of the samples ECAPed at 473 K or 673 K.

Sample	Microhardness (HV)	Ultimate tensile stress (MPa)	Total elongation (%)
Conform	102.6	286.4	66.7
473 K-4 P	162.9	333.2	24.7
473 K-16 P	201.2	583.4	37.9
673 K-4 P	150.9	365.5	22.5
673 K-16 P	168.6	461.1	26.9

SEM observation was done to clarify the failure mechanisms in the ECAPed Cu-Mg alloy. Figure 7 presents SEM morphologies of the fracture surfaces of the alloy subjected to 16 passes ECAP at two different temperatures. Some round and equiaxed dimples are observed on the fracture surface of the 16-pass sample ECAPed at 637 K (Figure 7a), which is a typical ductile fracture. This kind of fracture occurs due to microvoid formation and coalescence [17]. Some elongated dimples are seen at the fracture surface of the 16-pass sample ECAPed at 437 K (Figure 7b). This result occurs due to internal shearing between voids and seems to be governed by a simple shear deformation. The difference of dimple pattern between the two ECAPed samples may stem from the deformation mechanism change from dislocations slip to GBS. This phenomenon is attributed to the extremely small volume of the ultrafine grains with large angle boundary, and these grains are more inclined to form 45° slip plane under the applied stress (corresponding to Schmid's law). Therefore, GBS is much easier to happen than dislocation slip for the UFG Cu-Mg alloy.

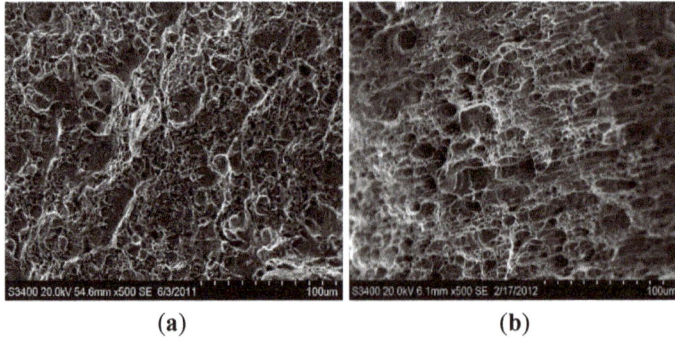

(a)   (b)

Figure 7. SEM fracture surface of the samples subjected to 16 passes ECAP at (**a**) 637 K and (**b**) 473 K.

3.4. Conductivity

The conductivity of the Conformed alloy after different ECAP passes were evaluated and is shown in Figure 8. It can be found that the conductivity of the alloy decreases when increasing the ECAP passes. The decrease of conductivity is attributed to the increase of grain boundaries, dislocations and large-angle grain boundaries caused by grain refinement. As is well known, grain boundaries and dislocations can increase the scattering of conducting electrons, leading to the increase of the electrical resistivity of the metal [8]. In addition, the large-angle grain boundaries have a large effect on the scattering of conducting electrons, greater than that of the low-angle grain boundaries [19].

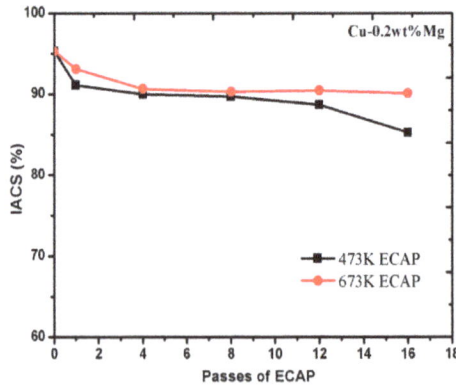

Figure 8. Conductivity of Cu-0.2wt%Mg alloy variation with the passes of ECAP at 473 K and 673 K.

Compared with the ECAPed samples at 673 K, the conductivity of the ECAPed samples at 473 K are obviously lower. This phenomenon should be caused by more grain boundaries (including large-angle grain boundaries) and dislocation multiplication during the ECAP process executed in lower temperature. Compared with the product of Cu-0.4wt%Mg contact wires fabricated by Conform + drawing process, the Cu-0.2wt%Mg alloy after Conform + 16-pass ECAP at 473 K has the higher conductivity (84.5% IACS). This good result should be attributed to the lower dislocation density and lower lattice strain after multi-pass ECAP processing.

Thus, it can be seen that grain refinement via multi-pass ECAP processing can endow the Cu-0.2wt%Mg alloy with superior strength and good conductivity characteristics, which are advantageous to high-speed electrification railway systems. This new technique can be easily

Metals **2014**, *4*, 586–596

integrated into the manufacturing processes of Cu-Mg alloy contact wire (*i.e.*, after the Conform process), of which successful application makes the trains safe at higher speeds.

4. Conclusions

(1) Multi-pass ECAP processing, compared with cold drawing, improves grain refinement effect of the Conformed Cu-0.2wt%Mg alloy. More passes of ECAP leads to finer grains, and the grain size of the 16-pass sample ECAPed at 473 K is about 200 nm.

(2) Compared with the as-achieved sample in Conform state, the ones after multi-pass ECAP exhibit much higher strength with adequate ductility. The ECAPed samples for 16 passes have much higher strength and better ductility than those for less passes.

(3) With increasing the ECAP pass, hardness and strengthen of the ECAPed samples increased obviously but the conductivity decreased gradually. However, the conductivity of the Cu-0.2wt%Mg alloy after Conform plus ECAP is still much higher than that of the current Cu-0.4wt%Mg product processed by Conform plus cold drawing.

(4) Conform plus ECAP provides a simple and effective procedure to obtain high strength and good conductivity Cu-Mg alloy, in comparison with current Conform plus cold drawing. After Conform plus ECAP for 16 passes at 473 K, Cu-0.2wt%Mg alloy exhibits superior tensile strength (583.4 MPa), adequate total elongation (37.9%), good conductivity (84.5% IACS), and high hardness (201.2 HV).

Acknowledgments: This work was sponsored by Qing Lan Project, National Natural Science Foundation of China (Grant No. 51141002) and Jiangsu Provincial Natural Science Foundation of China (No. BK20140856).

Author Contributions: The work presented here was carried out in collaboration between all authors. A. Ma, C. Zhu and J. Jiang defined the research theme. A. Ma, C. Zhu and D. Song designed methods and experiments, carried out the laboratory experiments, analyzed the data, interpreted the results and wrote the paper. J. Chen, S. Ni and Q. He co-designed experiments, discussed analyses and interpretation. All authors have contributed to, seen and approved the manuscript.

The author hopes that this paper can make its due contribution to successful application of high-strength and high-conductivity Cu-Mg alloy contact wire.

Conflicts of Interest: The authors declare no conflict of interest.

References

1.   Lu, L.; Shen, Y.; Chen, X.; Qian, L.; Lu, K. Ultrahigh strength and high electrical conductivity in copper. *Science* **2004**, *304*, 422–426. [CrossRef] [PubMed]

2.   Han, K.; Walsh, R.P.; Ishmaku, A.; Toplosky, V.; Brandao, L.; Embury, J.D. High strength and high electrical conductivity bulk Cu. *Pilos. Mag. A* **2004**, *84*, 3705–3716.

3.   Wei, K.X.; Wei, W.; Wang, F.; Du, Q.B.; Alexandrov, I.V.; Hu, J. Microstructure, mechanical properties and electrical conductivity of industrial Cu-0.5% Cr alloy processed by severe plastic deformation. *Mater. Sci. Eng. A* **2011**, *528*, 1478–1484. [CrossRef]

4.   Sakai, Y.; Schneider-Muntau, H.J. Ultra-high strength, high conductivity Cu-Ag alloy wires. *Acta Mater.* **1997**, *45*, 1017–1023. [CrossRef]

5.   Saarivirta, M.J. High conductivity copper-rich Cu-Zr alloys. *Trans. Metall. Soc. AIME* **1960**, *218*, 431–437.

6.   Embury, J.D.; Han, K. Conductor materials for high field magnets. *Curr. Opin. Solid State Mater. Sci.* **1998**, *3*, 304–308. [CrossRef]

7.   Lu, K.; Lu, L.; Suresh, S. Strengthening materials by engineering coherent internal boundaries at the nanoscale. *Science* **2009**, *324*, 349–352. [CrossRef] [PubMed]

8.   Reiss, G.; Vancea, J.; Hoffmann, H. Grain-boundary resistance in polycrystalline metals. *Phys. Rev. Lett.* **1986**, *56*, 2100–2103. [CrossRef] [PubMed]

9.   Fernee, H.; Nairn, J.; Atrens, A. Cold worked Cu-Fe-Cr alloys. *J. Mater. Sci.* **2001**, *36*, 5497–5510. [CrossRef]

10. Valiev, R.Z.; Islamgaliev, R.K.; Alexandrov, I.V. Bulk nanostructured materials from severe plastic deformation. *Prog. Mater. Sci.* **2000**, *45*, 103–189. [CrossRef]

11. Valiev, R.Z.; Langdon, T.G. Principles of equal-channel angular pressing as a processing tool for grain refinement. *Prog. Mater. Sci.* **2006**, *51*, 881–981. [CrossRef]

12. Torre, F.D.; Lapovok, R.; Sandlin, J.; Thomson, P.F.; Davies, C.H.J.; Pereloma, E.V. Microstructures and properties of copper processed by equal channel angular extrusion for 1–16 passes. *Acta Mater.* **2004**, *52*, 4819–4832. [CrossRef]

13. Iwahashi, Y.; Wang, J.; Horita, Z.; Nemoto, M.; Langdon, T.G. Principles of equal-channel angular pressing for the processing of ultra-fine grained materials. *Script. Mater.* **1996**, *35*, 143–146. [CrossRef]

14. Segal, V.M. Materials processing by simple shear. *Mater. Sci. Eng. A* **1995**, *197*, 157–164. [CrossRef]

15. Yamashita, A.; Yamaguchi, D.; Horita, Z.; Langdon, T.G. Influence of pressing temperature on microstructural development in equal-channel angular pressing. *Mater. Sci. Eng. A* **2000**, *287*, 100–106. [CrossRef]

16. Huang, J.Y.; Zhu, Y.T.; Jiang, H.; Lowe, T.C. Microstructure and dislocation configurations in nanostructured Cu processed by repetitive corrugation and straightening. *Acta Mater.* **2001**, *49*, 1497–1505. [CrossRef]

17. Shaarbaf, M.; Toroghinejad, M.R. Nano-grained copper strip produced by accumulative roll bonding process. *Mater. Sci. Eng. A* **2008**, *473*, 28–33. [CrossRef]

18. Habibi, A.; Ketabchi, M.; Eskandarzade, M. Nano-grained pure copper with high-strength and high-conductivity produced by equal channel angular rolling process. *J. Mater. Process. Technol.* **2011**, *6*, 1085–1090. [CrossRef]

19. Dannenberg, R.; King, A.H. Behavior of grain boundary resistivity in metals predicted by a two-dimensional model. *J. Appl. Phys.* **2000**, *88*, 2623–2633. [CrossRef]

MDPI AG

St. Alban-Anlage 66

4052 Basel, Switzerland

Tel. +41 61 683 77 34

Fax +41 61 302 89 18

http://www.mdpi.com

Metals Editorial Office

E-mail: metals@mdpi.com

http://www.mdpi.com/journal/metals

www.ingramcontent.com/pod-product-compliance
Lightning Source LLC
Chambersburg PA
CBHW051854210326
41597CB00033B/5899